More JOY of MATHEMATICS

Exploring Mathematics All Around You

by Theoni Pappas

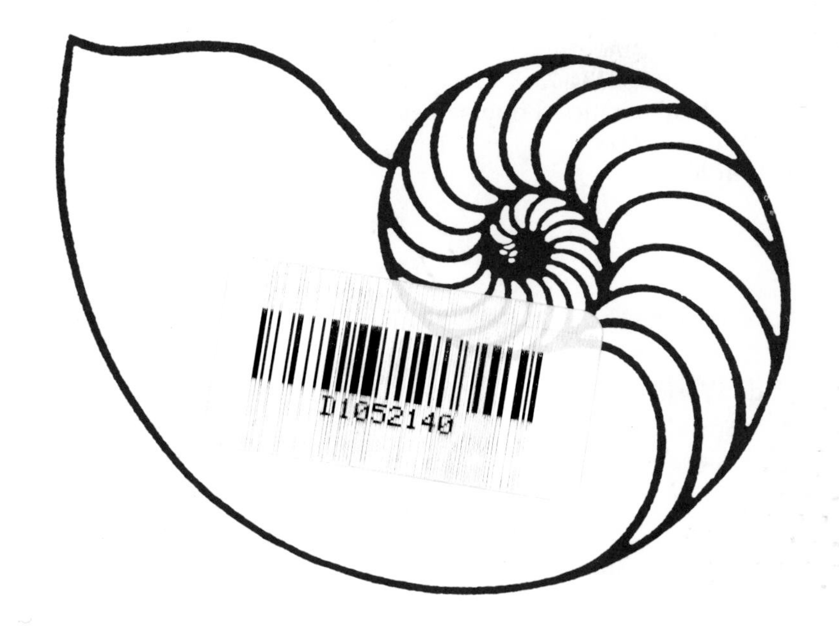

nature • science • music • architecture • philosophy • history • literature

WIDE WORLD PUBLISHING/TETRA

Library of Congress Cataloging–in–Publication Data

Pappas, Theoni.
More joy of mathematics: exploring mathematical insights & concepts / by Theoni Pappas.
p. cm.
Includes index.
ISBN 0-933174-73-X (pbk.) : $10.95
1. Mathematics--Popular works. I. Title.
QA93.P37 1991
510--dc20 91-11295
CIP

Printed in the United States of America

Wide World Publishing/Tetra
P.O. Box 476
San Carlos, CA 94070

1st Printing March 1991
2nd Printing August 1991

INTRODUCTION

Mathematics is more than doing calculations, more than solving equations, more than proving theorems, more than doing algebra, geometry or calculus, more than a way of thinking. Mathematics is the design of a snowflake, the curve of a palm frond, the shape of a building, the joy of a game, the frustration of a puzzle, the crest of a wave, the spiral of a spider's web. It is ancient and yet new. Mathematics is linked to so many ideas and aspects of the universe.

In mathematics one finds numbers, abstract ideas, concepts that appear cold and dull to the untrained eye. However, when mathematics is viewed as a collage, its creativity and beauty emerge. I ask you to stand back and gaze upon mathematics with open eyes and mind. Every time I see one of its treasures, I am made aware of the beauty of mathematics. With *More Joy of Mathematics* I hope you will marvel at its realm and scope and experience the joy there can be.

The topics in this book and in *The Joy of Mathematics* give but a glimpse of some intriguing idea, concept, puzzle, historical note, game, etc. I hope that they pique your curiosity to seek out and explore additional information.

ACKNOWLEDGEMENTS

Special thanks to my parents and grandmother for their love, support and encouragement, and to my teachers for expressing their enthusiasm for their subjects.

Special appreciation

- to the makers of mathematics from ancient to modern times,
- to the writers of the hundreds of books and articles I have been fortunate to find over the years,
- to Sam Loyd, Henry Dudeney and the other puzzlists for the hours of puzzling and frustrating enjoyment their works continually provide,
- to Martin Gardner, my present day mathematical guru, whose works continually introduce young and old laypeople and math enthusiasts to the excitement of mathematics,
- to Elvira Monroe who helped me learn to popularize my approach to mathematics,
- to Mía Monroe whose perceptive eyes and mind were invaluable in refining the pages of this book

Lastly,

- to the scores of readers who have written with encouragement, corrections, ideas and criticisms.

TABLE OF CONTENTS

TABLE OF CONTENTS

TABLE OF CONTENTS

TABLE OF CONTENTS

"The mathematician does not study
pure mathematics because it is
useful; he studies it because
he delights in it and he delights
in it because it is beautiful.

—Henri Poincaré

the mathematics of ocean waves

Ocean waves—magnificent cresting forms—seem to personify creatures of the sea as they swell, roll and break on shore. Sophisticated mathematical equations and properties have been developed over the centuries to describe them. To understand and explore some of the mathematics that is used to explain their varying shapes, sizes, forms and idiosyncrasies, a look at the general nature of waves is needed.

Consider two people creating waves by undulating a rope between them. Imagine the wave traveling through the rope. The rope is not moving closer to either person. What is being transported between the two people is energy. Thus, waves are motions that carry energy over a medium. In this case the medium is a rope. It may be water (ocean waves), the earth (seismic earthquake waves), electromagnetic field (radio waves), air (sound waves). Waves are created when the medium is disturbed or agitated in some way.

Ocean waves are created when either the wind, an earthquake, an object (such as a moving boat) and/or gravitational pull of the moon and the sun (causing tides) disturb the water (the medium). Ocean waves travel on the surface of the water. These undulating forms have a somewhat random quality to them when there are multiple disturbances.

Much study was generated in the 1800's on the mathematics of ocean waves. Observations at sea and in controlled laboratory experiments helped scientists arrive at interesting conclusions. It started in Czechoslovakia in 1802, when Franz Gertner formulated the initial wave theory. In his observations he wrote how water particles in a wave move in circles. The water in the crest* of the wave moves in the direction of the wave and that in the trough* (the lowest point of the wave) moves in the opposite direction. On the surface, each water particle moves in a circular orbit before returning to its original position. This circle was

found to have a diameter equal to the height* of the wave. There are circles being generated by the particles throughout the depth of the water. But the deeper the particle the smaller its circle. In fact, it was found that at a depth of 1/9 the wave length*, the diameter of the circular orbit is about half that of the circular orbit of a surface particle.

Since waves are tied into these circling particles and since sinusoidal* and cycloid* shaped curves also depend on rotating circles, it is not surprising that these mathematical curves and their equations are used in the description of ocean waves. But it was discovered that ocean waves are not strictly sinusoidal or any other purely mathematical curve. The depth of the water, the intensity of the wind, the tides are only some of the variables that must be considered when describing waves. Today ocean waves are studied by applying the mathematics of probability and statistics. A large number of small waves are considered and predictions are formulated from the data collected.

Some other interesting mathematical properties of ocean waves are:

1) The wave length depends upon the period.

2) The wave height does not depend on the period or the length (there are some exceptions in which the influence of the period and the length are minute).

3) The wave will break when the angle of the crest exceeds 120°. When the wave breaks most of its energy is then consumed.

4) Another way to determine when the waves will break is to compare its height with its length. When this ratio is greater than 1/7, the wave will break.

---

*Definitions for—

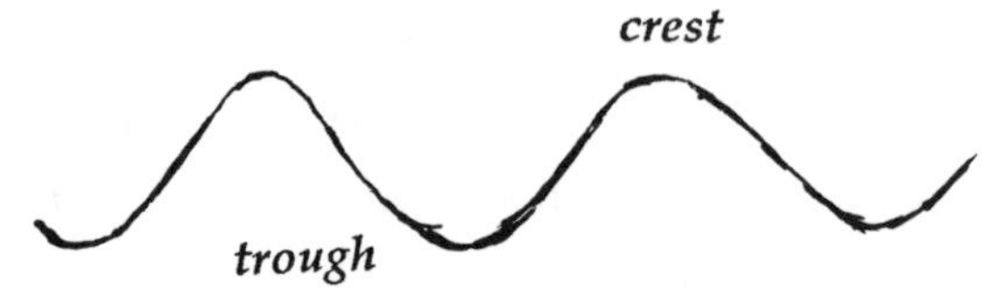

crest the highest point of the wave

trough the lowest point of the wave

wave height vertical distance from the crest to the trough

wave length horizontal distance between two consecutive crests

wave period the time in seconds it takes a wave crest to travel one wave length

sinusoidal curve is of the shape ∿∿∿ which is a periodic (regularly repeats its shape) trigonometric function.

cycloid curve is the curved which is the path of a point on a circle that is rolling along a straight line.

the tesseract uncubed

A cube drawn on a sheet of paper is sketched in perspective to imply its 3rd-dimension. This illustration shows one way to unfold a cube onto a 2-dimensional plane—

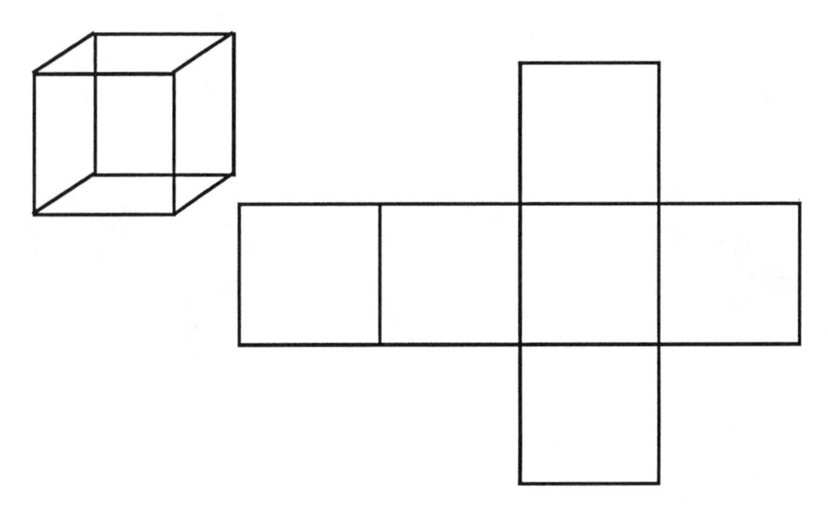

A hypercube or tesseract is a 4th-dimensional representation of a cube. Now apply a similar method to unfold a tesseract into the 3rd-dimensional space. The illustration shows that the tesseract or hypercube is composed of 8 cubes, 16 vertices, 24 squares, and 32 edges.

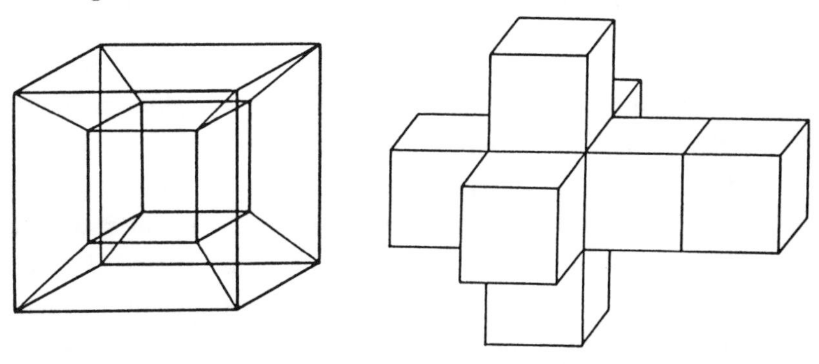

the tangram

One of the most popular puzzles of the 19th century was a Chinese puzzle called *the tangram.* Its popularity probably stems from how easy it is to use and understand the instructions. The tangram gives you the freedom to design your own figures, while challenging you to unravel how its seven pieces form particular designs. The simple instructions can deceive one into thinking the solution is easily arrived at. From the over 1600 designs some are easier than others to solve, but bear in mind some are *very* tricky and some are *paradoxical.*

Who would guess a tangram link between Napoleon Bonaparte, John Quincy Adams, Gustave Doré, Edgar Allan Poe, and Lewis Carroll? All were swept up in the tangram mania.

Although tangrams were being enjoyed for years prior, the earliest reference to the tangram appears in a Chinese book dated 1813, which was probably written during the reign of the Emperor Chia Ching (1796-1820). There are many theories on the origin of the name *tangram:*

(1) derived from the obsolete English word *trangram,* meaning *a puzzle* or *trinket;* (2) derived from a combination of the word *Tang* (Chinese dynasty) with suffix *–gram* (Greek meaning writing); (3) derived from the term *tanka,* families who in addition to ferrying for exporters also provided food, entertainment and did laundry. Some entertainment may have been in the form of the 7-piece Chinese puzzle and perhaps the

term *tanka game* evolved. Any of the theories is plausible.

Perhaps the most interesting and entertaining explanation of the origin of *the tangram* was literally created by America's famous puzzler, Sam Loyd in his book *The Eighth Book of Tan.* Loyd had a marvelous sense of humor and an affinity toward practical jokes. He wrote his book in 1903 when he was 61 years old . Why he waited so long is unknown, especially since his mother introduced him to the joy of solving and creating tangram puzzles and passed on to him the family's collection of two books* of tangram patterns. In his "history" he attributed the invention of the game to the god Tan, and he states *"According to the encyclopedia lore, the game of Tangrams is of very ancient origin, and has been played in China for upward of 4,000 years, somewhat in the nature of a national pastime...The seven Books of Tan were supposed to illustrate the creation of the world and the origin of species upon a plan which out-Darwins Darwin, the progress of the human race being traced through seven stages of development up to a mysterious spiritual state which is too lunatic for serious consideration. "* The Eighth Book of Tan was very convincing. In fact, some scholars were deceived into initially believing his false history until they had done extensive research. Nonetheless, Loyd's delightful, entertaining and

Loyd's Indians

superb puzzle book brings the tangram figures and designs to life with his running commentary.

As illustrated, the seven pieces of the tangram are composed of five similar right triangles, a square and a parallelogram. Challenge yourself by discovering how to assemble the seven pieces into a square. Then try to form two squares. Now try a rectangle. How about a parallelogram? The squaw and the Indian were Loyd's creations. Can you form them? The other seemingly paradoxical diagrams also appeared in his collection. Finally, create your own forms and figures.

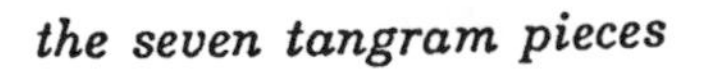

the seven tangram pieces

*John Singer, grandfather of painter John Singer Sargent, gave his sister Elizabeth Singer Loyd two books of tangram patterns he had collected, which Elizabeth passed on to her son Sam Loyd.

an elegant proof of the Pythagorean theorem

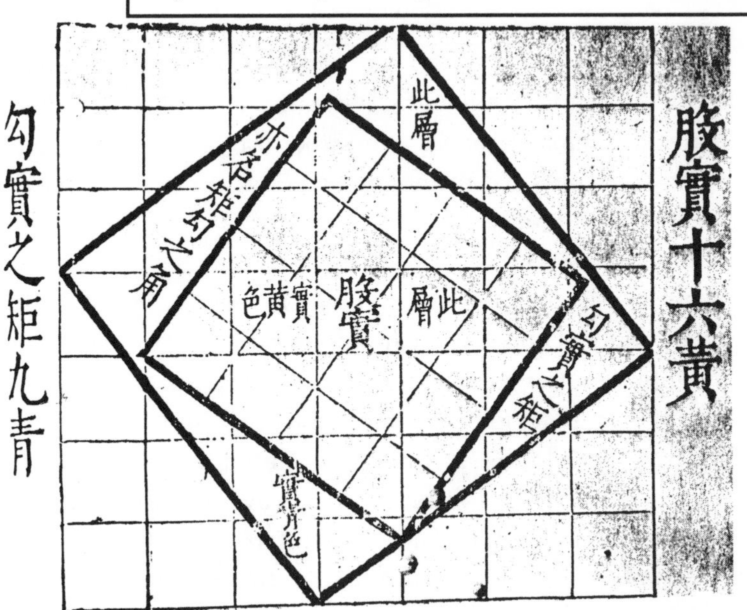

This diagram of the Pythagorean theorem was from the Chinese manuscript *The Chou Pei.* (There are conflicting beliefs as to the date of this work. Some estimate it as early as 1200 B. C. , while others set it at 100 A. D.) The work primarily deals with astronomical calculations and related mathematics.

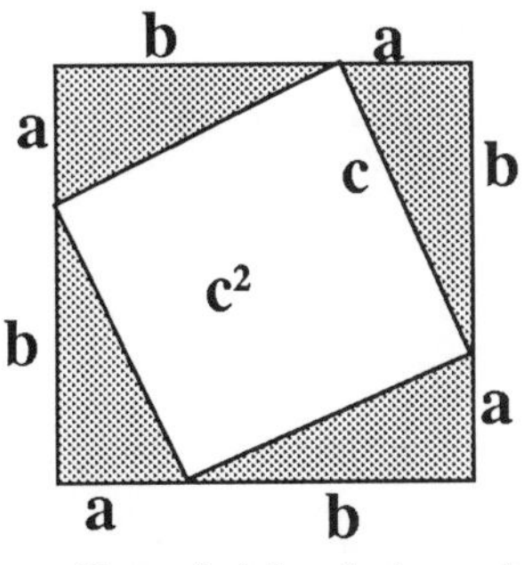

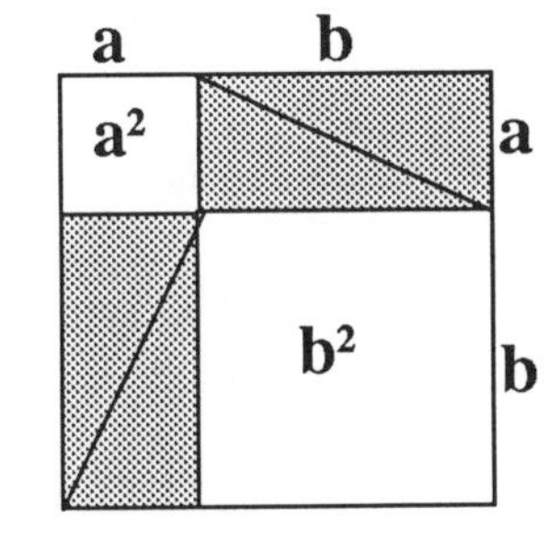

The unshaded region's area for this square = *The unshaded region's area for this square*

When this diagram is studied and rearranged, the proof of the Pythagorean theorem appears.

perplexing infinity

Infinity is an intriguing concept. It's a number, a quantity, represented by the symbol ∞ (invented by John Wallis in 1655) that describes an endless amount. Without the concept of infinity so many mathematical ideas would be lost. In fact, the development of calculus and the notion of a limit are inseparably tied to the notion of infinity.

Many perplexing ideas result when one begins to consider properties and areas where infinity occurs. For example, look at the infinite set of natural numbers, N={1,2,3,4,. . . }, which can be put into a one-to-one correspondence with the set of perfect squares, S={1,4,9,16, . . . },. Both are infinite sets. Yet consecutive elements of set N are always one unit apart, while consecutive elements of set S get farther and farther apart. Although it does not seem possible these sets have the same number of elements because every and any natural number of set N, call it k, has a matching element in set S, which is obtained by squaring k, k^2. So there is never an element of N that does not correspond to one of S or vice versa. The same holds true with the set of cubes, C={1,8,27,64,. . . } and so forth.

Infinity, in fact, is the culprit for so many seemingly paradoxical ideas.

1) An infinite amount need not take up an infinite amount of room. For example, there are an infinite number of points on segment AB, ________, yet AB is finite in length.

2) The sum of any infinite set of numbers need not be an infinite amount.
For example, $1/2 + 1/4 + 1/8 + 1/16 + . . . + (1/2)^n + ... = 1$

Proof

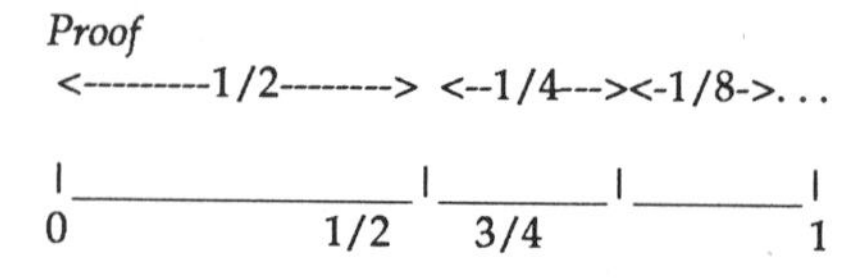

We see this sum will never pass 1.

This pumpkin patch in Half Moon Bay, California , has an infinite mood.

3) An object of finite length can be matched with an object of infinite length. The diagram below illustrates that the points of a semicircle (which has finite length) can be put in a one-to-one correspondence with the points of a line (which has infinite length).

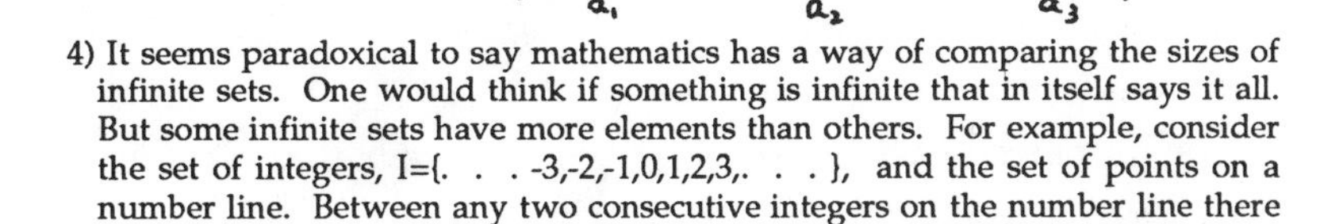

4) It seems paradoxical to say mathematics has a way of comparing the sizes of infinite sets. One would think if something is infinite that in itself says it all. But some infinite sets have more elements than others. For example, consider the set of integers, I={. . . -3,-2,-1,0,1,2,3,. . . }, and the set of points on a number line. Between any two consecutive integers on the number line there are yet an infinite number of points, and thus the line has more elements than set I.

5) There are occurrences of infinity and its paradoxes in nature. For example, the snowflake curve is formed with the astonishing characteristic that its area is finite while its perimeter is infinite.

Infinity has intrigued, baffled, astounded and challenged minds for centuries. Zeno's paradoxes date back to 5th century B. C. The four famous ones —*the Dichotomy, Achilles and the tortoise, the Arrow,* and *the Stade* — always provide interesting discussions and thoughts. And today the infinite nature of space-filling curves, the generation of fractals, the infinity of time, the finite or infinite nature of space, the continual search for a larger prime number, transfinite numbers for describing infinite sets, and others, *ad infinitum* create the atmosphere for inquiry and research in pursuit of infinity.

a twist to squaring the circle

Squaring the circle—constructing a square with the same area of a given circle using only a compass and straightedge— is considered one of the three famous construction problems of antiquity. It stimulated mathematical thought for over 2000 years, but it was only in 1882 that it was proven impossible.* In 1925 Alfred Tarsk removed the compass and straightedge restriction and proposed cutting up the circle into pieces and rearranging them into a square of the same area.

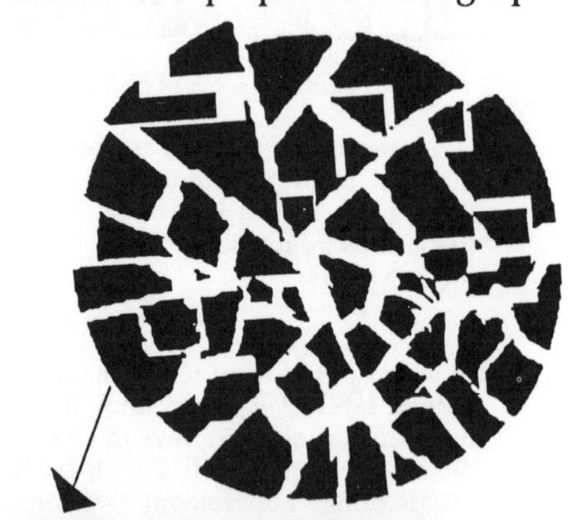

In 1989, Mathematician Miklós Laczkovich of Eötvös Loránd University in Budapest, Hungary has apparently proved that it is possible! His proof is now being scrutinized for errors or flaws. But none have yet to be detected. There are no gaps or overlapping pieces in the resulting square, and Laczkovich estimates it would require 10^{50} pieces.

* The proof uses the argument that a straightedge can be used to construct line segments whose equations are therefore linear. A compass on the other hand can construct circles and arcs whose equations are 2nd degree. When these types of equations are solved simultaneously by using linear combinations they produce at most second degree equations. But the equations derived in solving the three construction problems of antiquity by algebraic means involve transcendental numbers or cubic equations. Therefore, a compass and a straightedge alone cannot be used to derive these types of equations.

some tantalizing puzzles

bookworm puzzle

Suppose each volume of this set of books is 2. 5" thick, which includes the . 25" thick covers. A bookworm

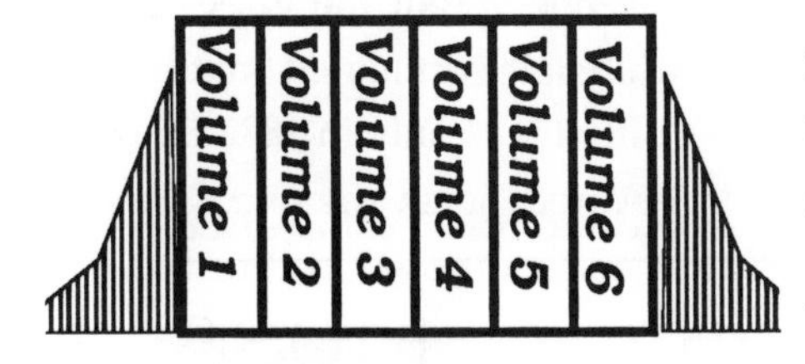

is 1" long. The bookworm eats in a straight horizontal line beginning at the front cover of the first volume and ending at the back cover of volume 6. How many inches did the bookworm travel?

the bamboo pile

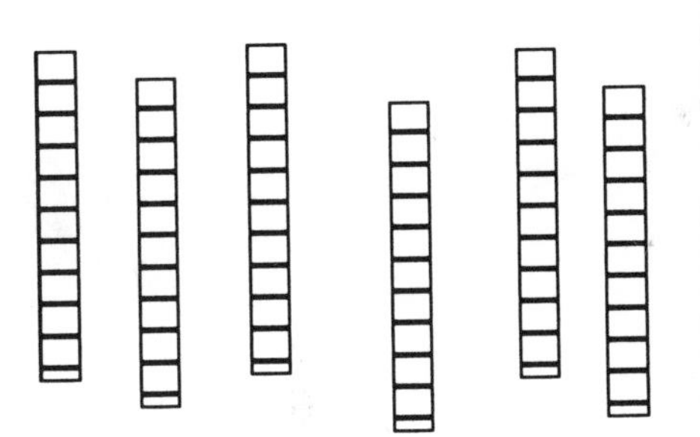

Pile six bamboo rods so that each of them touches each of the others. No breaking or bending allowed.

musical coins puzzle

The silver coins are to exchange places with the gold coins. The coins can move one space at a time either vertically, horizontally, or diagonally. Only one coin can occupy a square. Find the minimum number of moves.

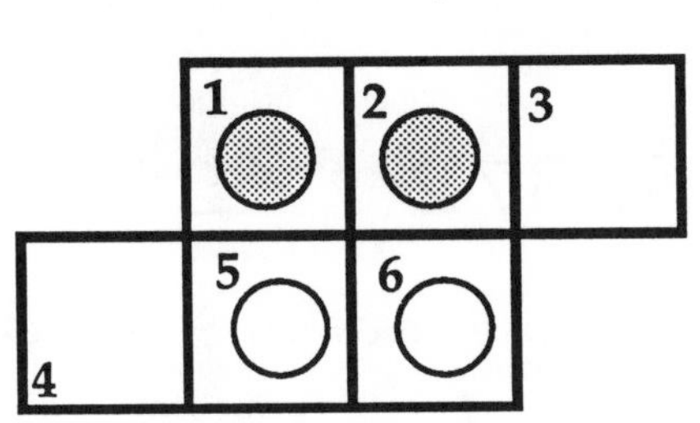

For solutions, see the appendix.

an amazing number property

Rearrange the digits in any whole number in any way you like. The difference between the first number and the new number will always be divisible by 9.

whole number choice	rearranged choice	difference
12563	23651	11088
		divided by 9=1232
87	78	9
		divide by 9=1
33333	33333	0
		divide by 9 =0
672636	666372	6264
		divided by 9=696

Fibonacci mania

1, 1, 2, 3, 5, 8, 13, 21, 34,...

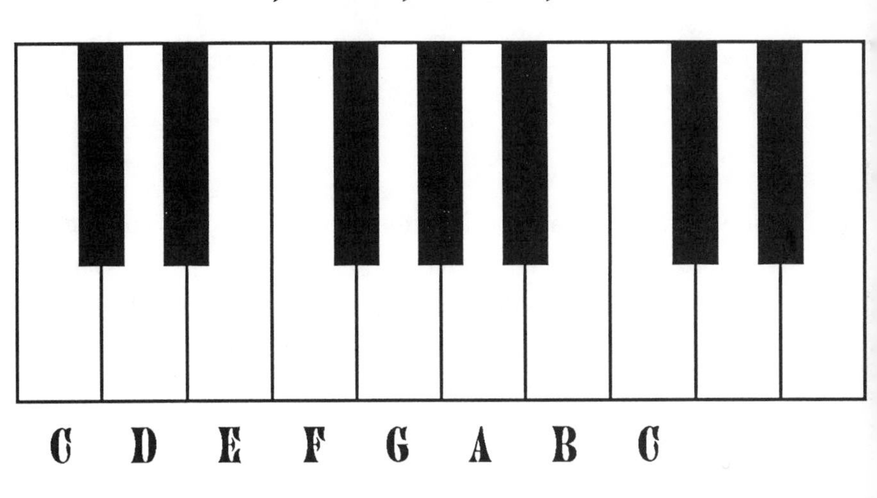

People have discovered so many areas where the numbers of Fibonacci sequence appears — including pine cones, pineapples, leaf arrangement, number of petals on certain flowers, pseudo-golden rectangles, golden mean connection, equiangular spirals — sometimes one wonders if we are overreacting to any time one finds *some* of the Fibonacci numbers present in a particular object. Is it just a coincidence that the white keys of a scale on the piano are *8* and the black keys number *5* ? Or is this another area where Fibonacci strikes again?

A mysterious formula for π

This formula gives a rapid way to calculate the value of π. Computer scientists using an adaptation of this formula calculated π to 17 million places.

$$\frac{1}{\pi} = 2\sqrt{2}\sum_{n=0}^{\infty} \frac{\left(\frac{1}{4}\right)_n \left(\frac{1}{2}\right)_n \left(\frac{3}{4}\right)_n}{(1)_n (1)_n n!}(1103+26390n)\left(\frac{1}{99}\right)^{4n+2}$$

Srinivasa Ramanujan was a mathematician who was completely enthralled by numbers. Born in 1888 in the city of Kumbakonam in southern India, his mathematical foundation was self taught. This fact may explain his original and unorthodox way of approaching problems. His life was devoted to an exploration of mathematical ideas, as evidenced by his voluminous formulas and pages of work. Not having the advantages of computers to test ideas, he manually carried out computations. His work could have been lost had he not in desperation written English mathematicians with some of his discoveries. Recognizing his genius, English mathematician Godfrey J. Hardy invited him to Cambridge. At the age of 25 he left his homeland and his wife to pursue his love and thirst for mathematics. At that time he was virtually ignorant of modern European mathematics. This accounted for gaps of knowledge in certain areas. His subsequent seven years yielded much work, learning and discoveries that are just now coming to light. He always worked feverishly regardless of his bouts of weakness caused by an unknown malady. His health continued to deteriorate, finally causing him

3.141592653589793238462643
383279502884197169399375
105820974944592307816406
286208998628034825342117
067982148086513282306647
093844609550582231725359
4081284811174502841027...

to return to India in 1919. He died at the early age of 32, in April of 1920. It was not until 1976, that his *Lost Notebook* was found. It was discovered in a box with letters and bills in the library of Trinity College in Cambridge by George Andrews, a Pennsylvania State University mathematician. Ramanujan's style of work was to use a slate and erase as he went along. When he arrived at a particular formula, he would write it in his notebooks. Thus the final form is there, but the intermediate steps are missing. In fact, his notebooks contain about 4,000 formulas and other work.

Mathematicians are studying, using and trying to prove his formulas. As mathematician Jonathan Borwein of Dalhousie University in Halifax Nova Scotia, remarks *"When he pulled extraordinary objects out of the air, they weren't just curiosities but they were the right things. They are elusive evidence of a theory that's lurking around somewhere that he never made explicit."*

hyperspace—a disappearing act of mathematics

Hyperspace, the world of the tesseract or hypercube, the hypersphere, and all other 4-dimensional objects and entities, has tantalized and taxed the imagination of mathematicians. Here is a world which we 3-dimensional creatures can only visit through our minds. Our knowledge of dimensions and how they progress logically implies that such a place must exist. Our understanding of hyperspace depends on our ability to draw an analogy between our 3-D world and the world of 2 dimensions.

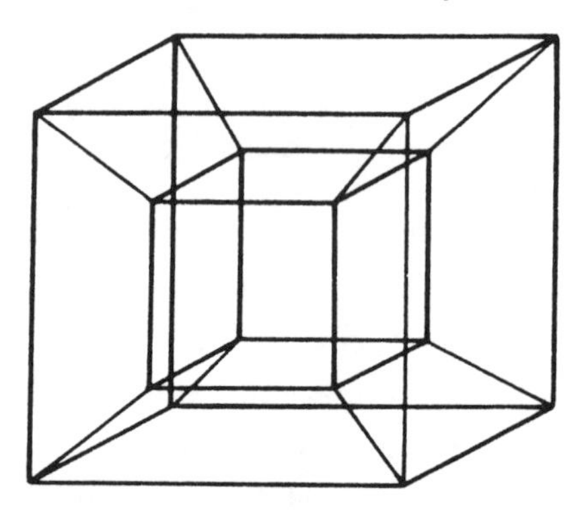

The tesseract is a 4th-dimensional representation of a cube. A cube drawn on paper is sketched in perspective to imply its 3rd-dimensional characteristic. A tesseract drawn on paper is a perspective of a perspective.

What kind of phenomena occur in hyperspace? What can hyperbeings do, and why do their acts initially seem so alien to us? A hyperbeing could effortlessly remove things before our very eyes, leaving us with the feeling that these objects simply disappeared. Consequently, the hyperbeing can see inside any 3-D object or life form, and if need be remove anything from inside. Three-dimensional knots simply fall apart in the hands of a hyperbeing. A pair of our gloves can be easily transformed into two left gloves or two right gloves. Such occurrences seem far fetched, but with mathematics they become very logical phenomena.

A hyperbeing enters our world by intersecting our space, just as a sphere would enter a 2-D world by intersecting a plane. What we would see of the hyperbeing would only be an impression (one of its facets) of the hyperbeing. The impression a sphere leaves when passing through a plane would be a series of circles to us. Just as the sphere has another depth to itself which is invisible to a creature in a 2-D world, the same is true for the hyperbeing and our 3-D world. We 3-D creatures can enter a 2-D world and remove any object by simply pulling it into our third dimension. Thus it would seem to disappear to any 2-D creature.

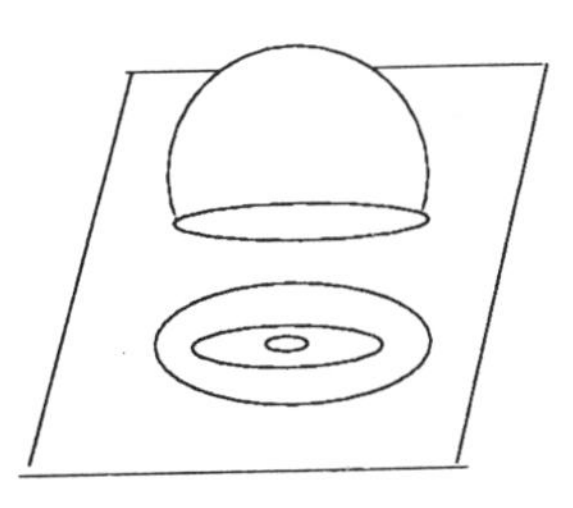

A sphere passing through a plane. It leaves circular impressions on the plane

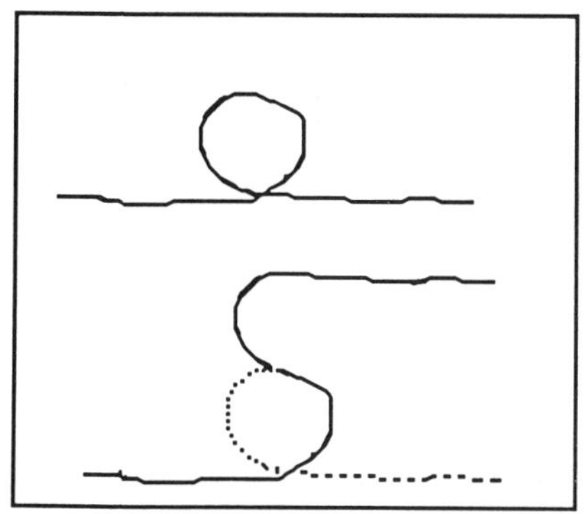

A knot in a 2-D world would be a flat loop. The diagram illustrates how a 2-D creatures has to undo a 2-D knot. 3-D creature could easily take apart a 2-D knot simply by lifting it into the 3rd dimension.

A 2-D creature's knot is effortlessly undone by a 3-D being because a knot in a 2-D world is no more than a loop. And by lifting the end of the loop up into the third dimension, the 2-D knot falls apart. This would happen to our 3-D knots if a hyperbeing were to take hold of it.

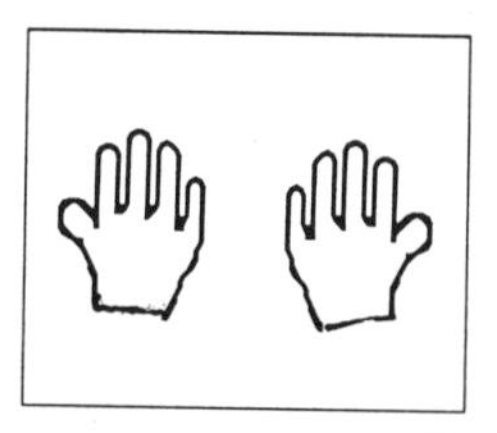

A right handed glove of a 2-D world could easily be changed into a left handed glove by flipping it over through the 3rd dimension.

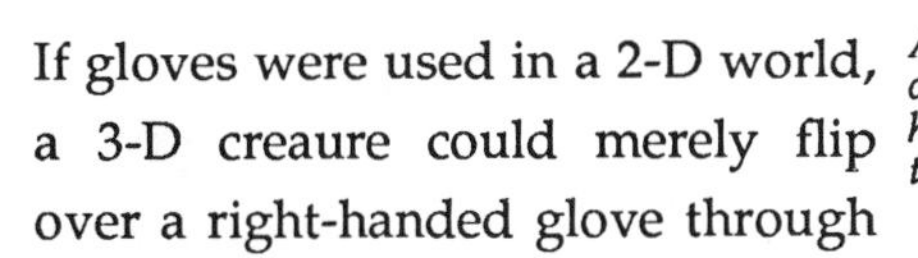

If gloves were used in a 2-D world, a 3-D creaure could merely flip over a right-handed glove through

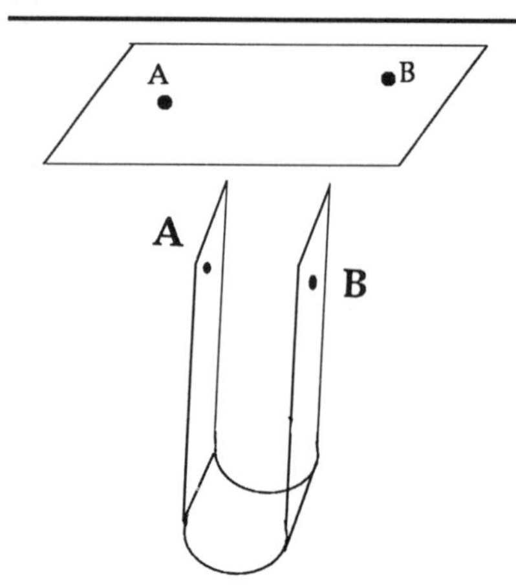

the third dimension, just as one would flip a pancake, and it would thus end up being a left-handed glove.

With a hyperbeing friend, travel would be very easy in a 3-D world. The hyperbeing would bend 3-space through the 4th dimension in the same way a 2-space (plane) could be bent by a 3-D creature to bring two points together.

A 3-D creature could enter any room, desk, closet of this 2-D house from above (the 3rd dimension), and could remove any object through the third dimension rather than walking though the hall or doors.

Hyperbeings would definitely seem like miracle workers in a 3-D world. What marvelous surgeons these hyperbeings would make. They could enter a 3-D being and remove or patch up a medical problem without ever making an incision. In a similar manner, we could enter any part of a 2-D world.

Regardless of whether you believe in higher dimensions, the mathematics behind it is very convincing.

the self-generating golden triangle

The golden triangle is an isosceles triangle with base angles 72° and the vertex angle 36°. When both of these base angles are bisected the two new triangles produced are also golden triangles. This process can be continued indefinitely up the legs of the original golden triangle, and an infinite number of golden triangles will appear as if they are unfolding.

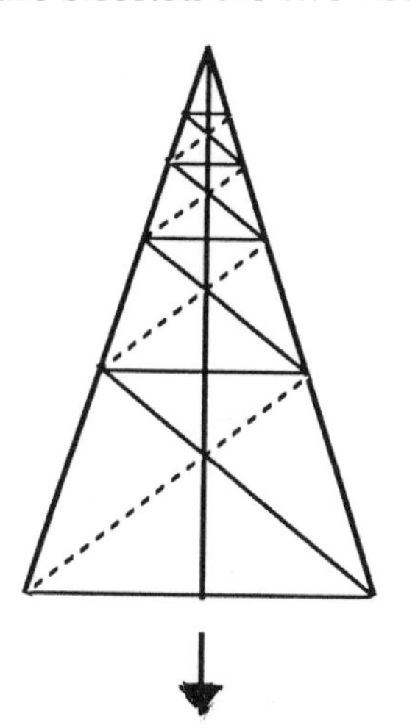

As this diagram shows, the golden triangle also produces the equiangular spiral and the golden ratio, $\phi = |AB|/|BC| \approx 1.618\ldots$

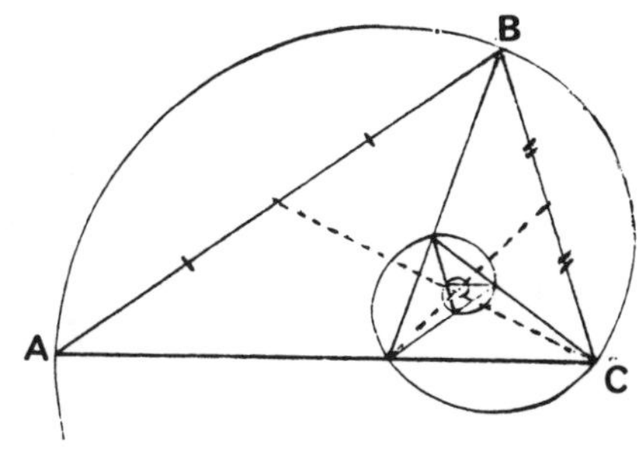

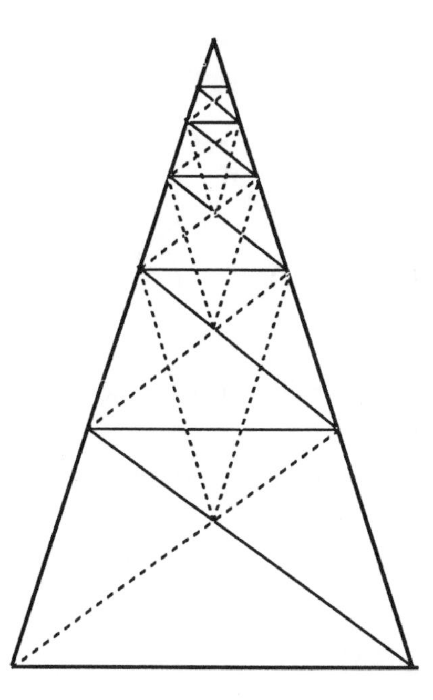

From these infinite climbing golden triangles one can also construct inside them an infinite number of climbing pentagrams. Note the five points of the pentagram are also golden triangles.

folding an ellipse

Many mathematical objects and ideas can be demonstrated by paper folding. Conic sections are no exceptions. Here is how to use paper folding to form an ellipse.

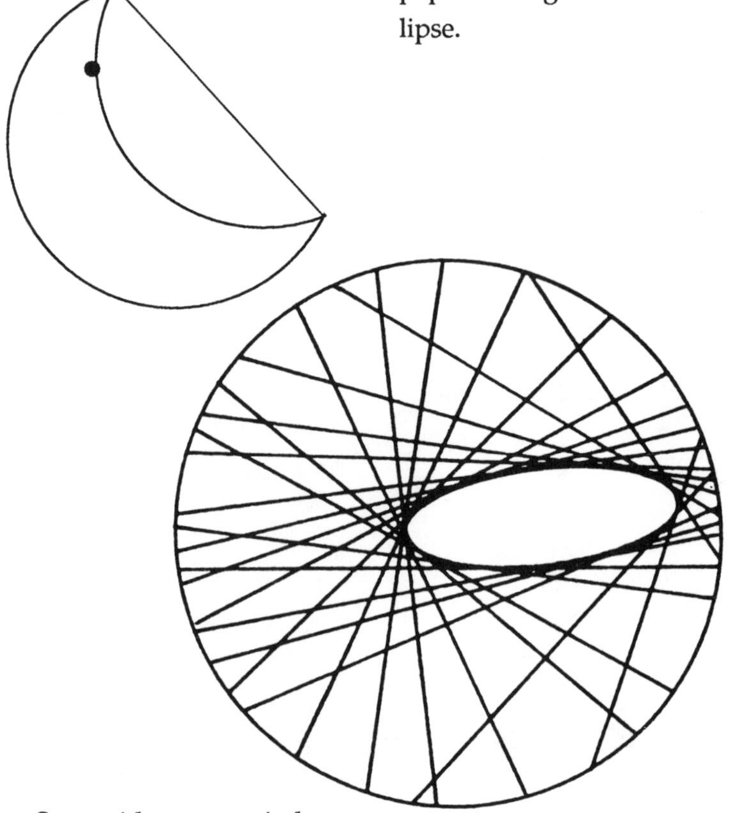

- Start with a paper circle.
- Select a point in the interior of the circle that is not the center. Mark its location with a dot.
- Fold and crease the circle so a point of its boundary lands on the dot.
- Continue the above process working your way around the circle's boundary.

Eventually the shape of the ellipse will form.

folding a hyperbola

Here's how to fold a hyperbola.

- Begin with a circle drawn on a sheet of paper.

- Select a point in the circle's exterior and mark its location with a dot.

- Fold a point of the circle's boundary to the dot as shown and crease the paper.

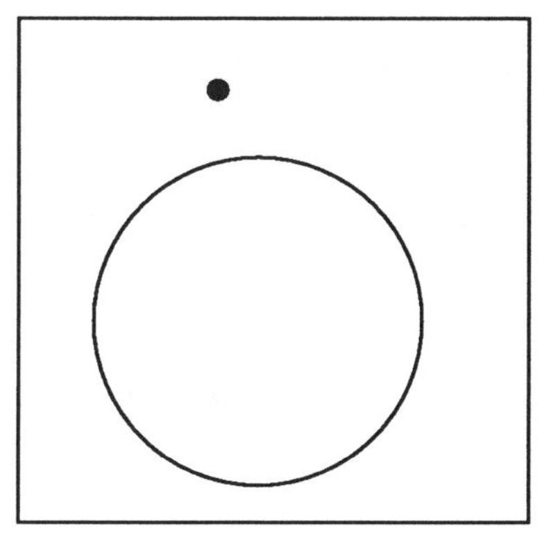

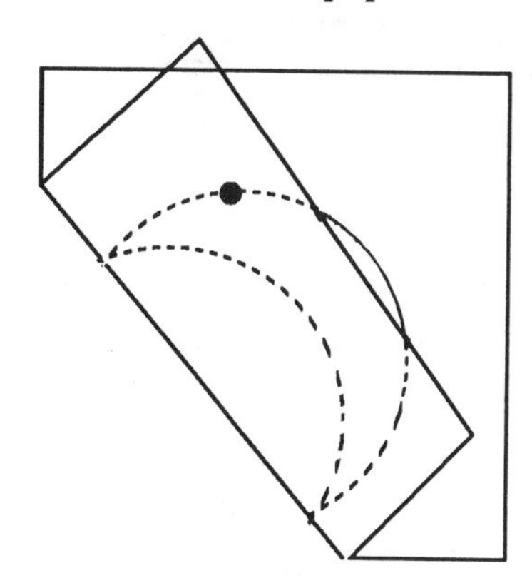

- Continue this process of folding and creasing around the entire boundary of the circle.

The hyperbola will eventually take form.

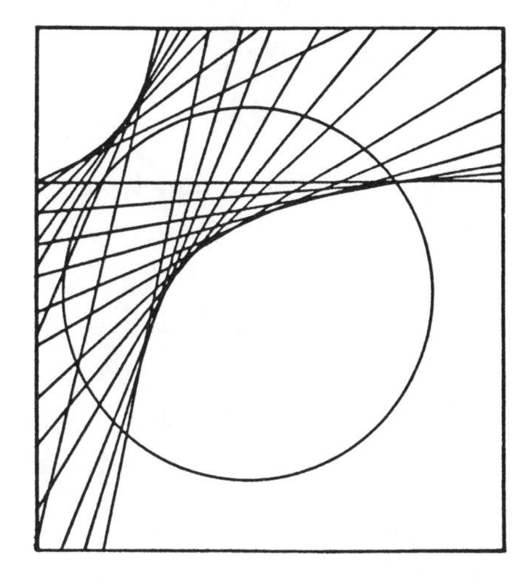

magic binary cards

The binary number system can be used to generate a stack of "magical" cards.

Card E

16	17	18	19
20	21	22	23
24	25	26	27
28	29	30	31

16's place

Card D

8	9	10	11
12	13	14	15
24	25	26	27
28	29	30	31

8's place

Card C

4	5	6	7
12	13	14	15
20	21	22	23
28	29	30	31

4's place

Card B

2	3	6	7
10	11	14	15
18	19	22	23
26	27	30	31

2's place

Card A

1	3	5	7
9	11	13	15
17	19	21	23
25	27	29	31

1's place

These five cards uniquely represent the numbers from 1 to 31. For example, 21 in base two is 10101, therefore 21 appears only on cards E, C and A. No two numbers have the same appearance on the cards because no two numbers have the same representation in base two. Thus, if someone says they are thinking of a number that appears on cards E, C and A—a quick mental computation of 16+4+1 gives 21*!

---

*In the binary system, 1 and 0 are the only digits used to write any number. Each card represents a place value in the binary system, for example, card E is the 16's place or 2^4. The number's binary representation indicates on which cards to place the number. Any place a 0 appears, the number is not placed on that card.

Pong hau k'i

In China this game is called Pong hau k'i, while in Korea it is called Ou-moul-ko-no. It seems like a very simple game because the playing board is so small, and only four playing pieces are involved. But don't be deceived!

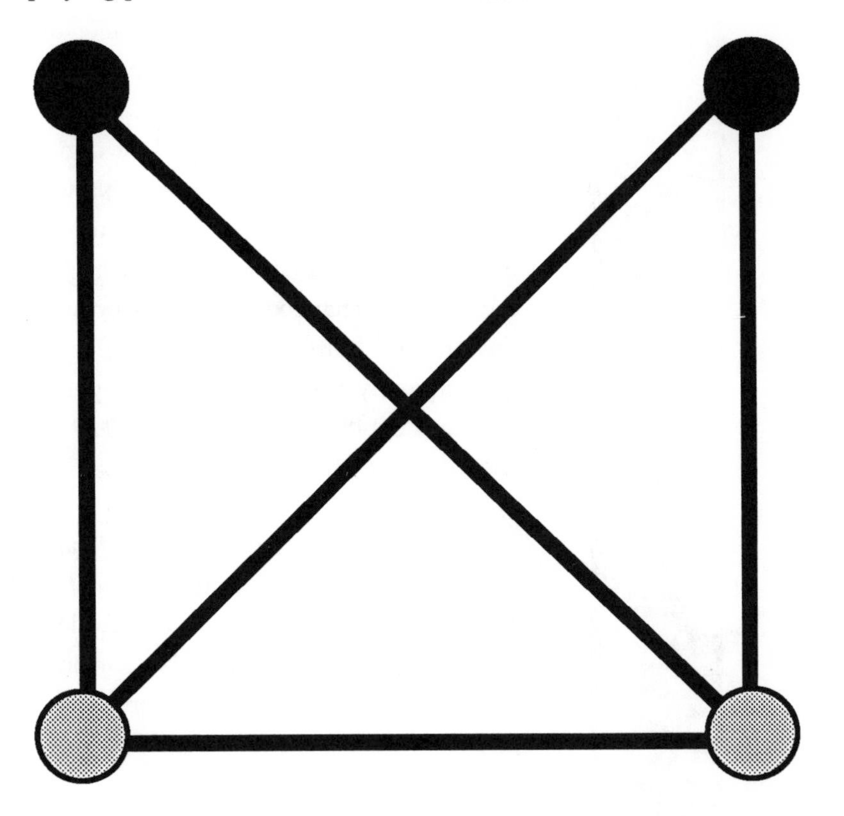

Procedure:
Each player has two playing pieces or special stones. They alternate placing them at the beginning. Once their four pieces have been placed, the players alternately move one piece at a time along any line to the next empty place. The object of the game is to block your opponent's stones so that he or she cannot move.

Egyptian cubits, palms & digits

Egyptian measuring tools initially were parts of the body. The *cubit* * was defined as the distance from the elbow to the end of the middle finger. Each cubit was divided into seven smaller units called *palms,* i. e. the width of one's hand. Each palm was subdivided into four *digits*—namely the four fingers or digits of the hand (excluding the thumb). Naturally no one had trouble carrying around these measuring instruments with them, but their lengths changed depending on whose anatomical parts were being used. Consequently the Egyptians developed two standard cubits, the *Royal Cubit≈20.59"* and the *short cubit≈17.72".*

The Egyptians constructed metal bars to correspond to a *Royal* and *short cubit* with subdivisions of *palms* and *digits* included on the bar. These bars can be considered predecessors to present day rulers.

*The term cubits was also used by the Greeks and Romans for their measuring units. A Greek cubit was approximately 18.22" and the Roman cubit was about 17.47".

Egyptian cubit measuring rod with subdivisions of palms and digits.

large numbers—the hidden solutions

In 1769 Leonard Euler, as a spin-off of *Fermat's Last Theorem**, hypothesized that $a^4+b^4+c^4=d^4$ had no possible positive integral solutions. Now 200 years later, using computers mathematicians have found integers which satisfy the equation and therefore contradict Euler's conjecture.

Noam D. Elkins of Harvard University discovered the first example, a=2,682,440; b=15,365,639; c=18,796,760; and d=20,516,673. Lately, Roger Frye of Massachusetts has apparently found the smallest positive integers that work — a=95,800; b=217,519; c=414,560; and d=422,560.

2,682,440
15,365,639
18,796,760
20,516,673

Mathematicians using computers contradict Euler

*See page 150 for additional information on *Fermat's last theorem*.

computer modeling

Today mathematicians are using the computer to visualize, create models and delve into once indescribable worlds. The old saying "changing before one's eyes" may not be so far fetched when referring to the innovations taking place in computer imagery. The changes in the last decade are mind boggling. The concept of the holodeck from *Star Trek, The Next Generation* may not be far off. Professions from many areas are discovering and using computer imagery—surgeons can use computer images to study surgical areas, architects can view their designs in completed forms with landscape and different lighting from any angle, environmentalists can predict outcomes of natural phenomena, pilots can experience all sorts of circumstances in jets without leaving the ground, musicians can create musical scores using the computer as an instrument, health officials can track and anticipate the spread of contagious diseases, cinematographers and artists can create a realistic fantasy scene without lifting a brush, and the list goes on. Mathematicians are working with graphics experts and computer scientists to launch new high tech methods of visualization. Complex methods are being devised with sophisticated computer programs to create multicolored realistic models to explore mathematical ideas never viewed outside the mathematician's mind. These models are helping solve problems and make conjectures in such areas as Knot Theory, soap bubbles, hyperspace, tessellating space, non-Euclidean models. Almost any problem or idea formulated by one's imagination can be shared more easily with others via computer modeling.

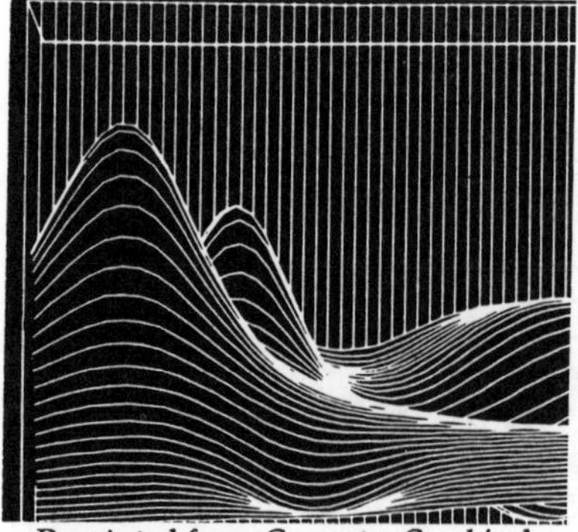

Reprinted from *Computer Graphics* by Melvin L. Prueitt. Courtesy of Dover Publications.

solitaire checkers puzzle

This puzzle has a variety of names including, *eight men in a boat* and *all-in-a-line.*

The white pieces must change places with the black pieces while following these rules:

1) pieces of the same color cannot jump one another.
2) move one piece one space or jump at a time

Find the least number of moves.

To make the game more challenging, increase the number of pieces.

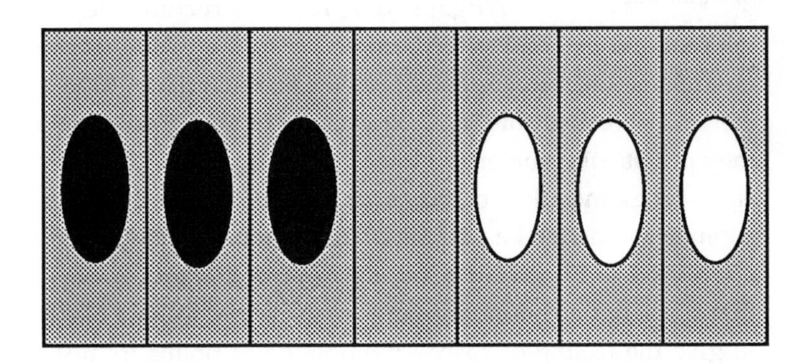

For the minimum number of moves, see the appendix.

the destruction of the box— architecture of Frank Lloyd Wright & the liberation of space

The work of Frank Lloyd Wright has a definite style, yet his structures are so diverse that the style does not lie in the similarities of his buildings, but rather in the philosophy that the structure projects. In his words, "Architecture is the scientific art of making structure express ideas. " His architecture has come to be called *organic architecture* — encompassing landscape, materials, methods, purpose and imagination in a special way.

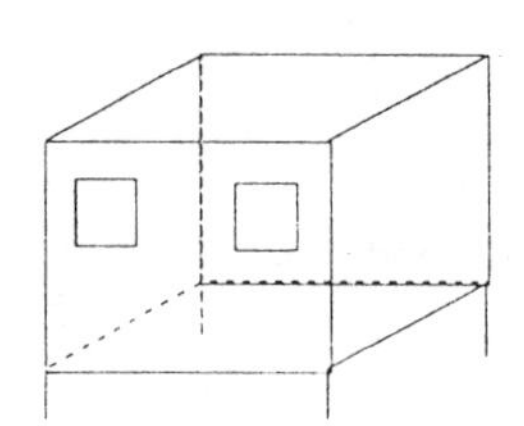

box architecture

Wright's designs of structures in which internal and external space become one had a profound impact on architecture. He designed buildings so that the outside came inside. He called this the *destruction of the box.* Dwellings, be they private or commercial, were viewed by Frank Lloyd Wright as a conglomeration of boxes or cubes. Space in Euclidean geometry is defined as the set of all points. Although a cube is often used in Euclidean geometry to represent space, we know that space has no boundaries or limits. Wright wanted his works to give the feeling of space—the flowing of points from the inside to the outside. Thinking along these lines, he discovered a way to eliminate the traditional box from his designs. He sought a change from the feeling of confinement and separation from the outside world which characterized box architecture. Wright realized that the potential of certain building materials had not been utilized. These materials — steel and glass — along with innovative design changes furnished the means to do away with the box, thereby allowing the merging of interior and exterior space.

Wright's designs did away with corners of the box by removing the supports from the corners and relocating them along the walls by using cantilevers.

Wright's Marin County Civic Center was one of his final designs. Marin County, CA.

Thus, the inhabitant's eyes were not bound or led to corners. Space was allowed to flow. By replacing post and beam construction with cantileverage supports, walls were no longer viewed as enclosing walls, but rather as independent and unattached. Any one of these walls could be modified by either shortening, extending or redividing.

Wright did not stop with freeing the horizontal plan, but also the vertical. He did away with cornice, and opened up the top to the sky. His designs eliminated the stacking and duplication of boxes. Instead, he used columns, and made them part of the ceiling, thereby creating a continuity of form. Now space inside and outside of a structure could move in all directions. The freedom of allowing space to move in and out of a structure is the essence of *organic architecture.* "So organic architecture is architecture in which you feel and see all this happen as a third dimension ... space alive by way of the third dimension." *

*Frank Lloyd Wright, *An American Architecture*, ed. Edgar Kaufman. New York: Bramhall House, 1955.

And the π goes on

It certainly seems knowing the value of π to four decimal places suits most people's everyday needs.

Spending 20 years doing calculations by hand, British mathematician William Shanks worked π to 707 decimal places—sadly he had made an error in the 528th place which went undetected until 1945.

π=3.14159265358979323846...

Why would one want to calculate π's value to millions of places, as is done today with supercomputers? Why this fascination for π's decimals?

- *It tests the capabilities of software and hardware of supercomputers.*
- *Calculation methods and ideas give rise to new ideas and concepts.*
- *Is π really patternless, does it contain endless variety of patterns?**
- *Do some of π's numbers appear more frequently than others, and therefore perhaps are not perfectly random?*

Perhaps mathematicians' fascination with π over the centuries can be likened to the drive that motivates mountain climbers to attempt an ascent.

---

*Some unusual frequencies found in π's decimal places are—*the first six digits of e occur 8 times thus far, the first 8 digits of √2 begin their appearance at 52,638th decimal, the first six digits of π (314159) appear at least 6 times among the first 10 million decimal places of π.*

tracking earthquakes mathematically

Earthquakes are tracked by studying the waves they produce. Three types of earthquake waves occur which arrive at a seismograph at different times. The P waves (5 miles per second—compressional longitudinal waves resembling sound waves) make rocks vibrate in the direction they travel. The S waves (3 miles per second—shake waves) vibrate rocks at right angles to their path. The L waves (2.5 miles per second) are surface waves that are confined to the earth's surface even as ocean waves are. Since these three waves travel at different speeds their distinct shapes appear on a seismograph at different intervals. First the P waves appear. When they die down, they are followed by the S waves. If the recording seismograph is located near the quake's epicenter, the S waves will arrive more quickly. The differing speeds of the P and S waves enable the distance traveled and the time of the quake to be determined. Since waves are reflected by different matter, this data also provides information on the earth's structure. Most earthquakes occur in the outer crust of the earth, and the absence of L waves would indicate an earthquake of deep origin.

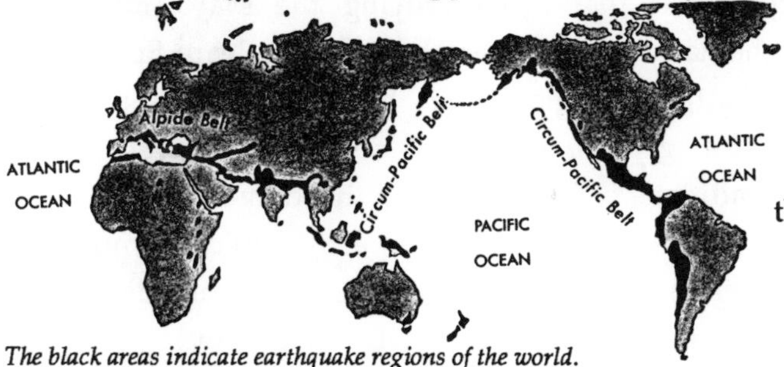

The black areas indicate earthquake regions of the world.

When an earthquake occurs, seismologists at various stations study the time intervals at which different types of waves reach them. Circles are drawn on a map indicating the earthquake's distances from the various stations. The earthquake's epicenter is located by studying the intersections of these circles.

Maya mathematics

The zealous fanaticism of the Inquisition and the greed of the Spanish conquerors resulted in the loss of most of the records and documentation of the Maya civilization.

The book *Relacion de las cosas de Yucatan* was overlooked for years in the library of the Spanish Royale Academy in Madrid, until in it was found in 1869 by the French priest, Charles Etienne Brasseur de Bourbourg. Fortunately, this book brought to light the existence of a civilization which its author, Friar Diego de Landa tried to destroy by burning the native books written in Maya glyphs. De Landa, in his zeal to spread Catholicism, felt that by destroying the Maya methods of writing, the natives would then be forced to learn Spanish, and thus the Catholic religion. His book describes some of the hieroglyphics (glyphs) that were destroyed as well as some cultural traditions. Consequently, the only examples of the Maya numerical notation appear in astronomical computations and computations of time through the Maya calendars. There are no writings linking their mathematics and architectural achievements—therefore any conclusions are only hypothetical.

There are theories, developed by D. Drucker, G. Kubler and H. Harleston which try to link architectural measurements of Mexican pyramids to Pythagorean numbers, π, e and the golden ratio.

The mathematics for which we *do* have documentation includes — the number system, place-value, zero and the impressive creation of their methods of keeping track of time.

Unfortunately, most of the writings of the Maya still remain undeciphered. We do know they counted by twenties. We have

A rendition of part of a page from Maya calendar records.

no evidence of the system they used on a daily basis, since their everyday practices and records were destroyed. We rely on what remains in their astronomy and calendar records. They developed a place-value system in which they could express very large numbers (in fact as large as they wanted). They also used a symbol to denote the absence of a unit, and therefore had come up with the concept of zero. Because religion and ceremonies played major roles in their lives and the control of their lives, astronomy and calendar records were extensive.

The calendars the Maya developed are more accurate than our present day Gregorian calendar or any other calendar used by other civilizations.

They developed three calendars—the **ceremonial calendar** (based on a 260 days cycle), the **solar calendar** (based on a 365+ day cycle), and the **Venus calendar** (based on 584 day cycle).

The Maya solar calendar was set up so that its error was only 1.98/10,000 of a day as compared to the 3.02/10,000 of a day for our Gregorian calendar. Astronomers from the city of Palenque

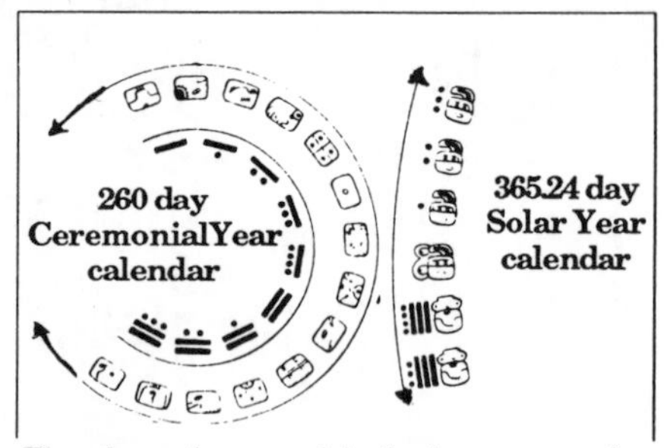

The solar and ceremonial calendars converged every 52 years—called the Calendar round.

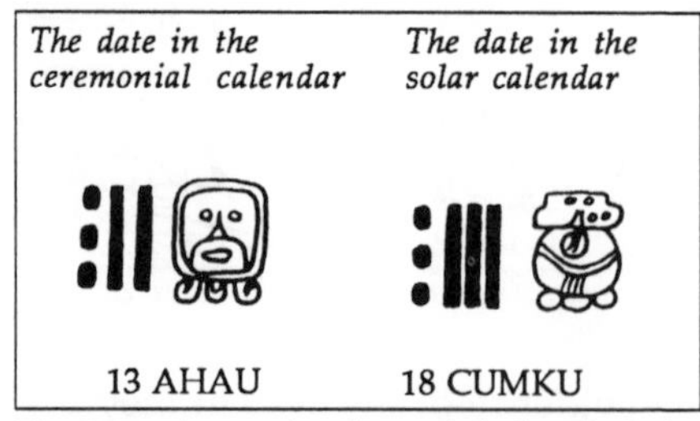

The date written using the symbols from both the ceremonial (sacred) calendar and the solar calendar.

calculated the average period of two successive full moons to be 2392 days for 81 lunar months which gives 29.5308642 days while modern astronomical calculations give 29.53059 days.

The Maya calendars were not strictly measures of the concept of time. They were personifications linking the names of their deities to dates. Each date was considered to carry a burden by a god (being auspicious or not) and was passed along cyclically. The alternating cycles of days, months and years were linked to endlessly recurring cycles.

The ceremonial calendar *(tzolkin)* consisted of 260 days. It was composed of twenty day cycles, each named after a god. Associated with each god's glyph was a number from 1 to 13. As a result each day of the 260 day calendar had a unique label composed of a god's glyph and a number which took 260 days before a repetition would occur. Consequently there was a 1-to-1 correspondence between the the gods' glyphs and the sequential repetition of the numbers 1 through 13.

god glyph A B C D E F G H I J K L M N O P Q R S T A B C D E F G...

numbers 1 2 3 4 5 6 7 8 9 10 11 12 13 1 2 3 4 5 6 7 8 9 10 11 12 13 1...

Somewhat analogous to astrology, the Maya associated each of these 260 days with characteristics that were deemed favorable or unfavorable days depending on the symbols. They believed the nature of the day you were born on characterized you and your entire life.

The solar calendar called *haab* was composed of 18 months of 20 days each. The five remaining days of the 365 day solar calendar were named *Uayeb,* and were considered lifeless and useless days. Individuals did very little during this five day period and their actions, during this time, were restricted by their religious beliefs.

The solar and ceremonial calendars were used concurrently — consequently a day would take both dates into account. Considering both systems as composing single manner of dating, the entire system would repeat every 52 years. The combination of calendars was called the *Calendar Round.* 365x260/5=18,980 days=52 solar years or 73 ceremonial years (i.e. 18,980 is the least common multiple for 260 and 365). This meant for a span of 52 years each day had a unique label.

Stir into this their Venus calendar, which is based on the number of days Venus is seen on the opposite side of the sun than the earth. Thus, each day for thousands of years had a unique name or label described using the names from the three calendars. Venus' synodical revolution varies between 580-588 days over a span of five revolutions. This averages to 583.92 days, which the Maya made 584 days. They calculated corrections for this discrepancy to be 24 days for 301 revolutions which made the error only one day in 6000 years!

Naturally tied into all these calendars and names of dates is an intriguing Maya lore and religion, which I urge the reader to explore. According to the Maya calendars, our present cycle began August 12, 3113 B.C. and will end with the destruction of the world on December 24, 2011 A.D.

The Maya numeration system was very likely influenced by their calendars. Because they used a modified base 20 place value system, they wrote their numbers vertically using only three symbols — • for one, ▬ for five and ⬬ for zero. *But their place value did not follow a strict base 20, i.e.* $1=20^0$, $20=20^1$, $400=20^2$, $8000=20^3$, $160{,}000=20^4$..., *but deviated after the third place, i.e.* $1=20^1$, $20=20^1$, $360=18x20^1$, $7200=18x20^2$, $144{,}000=18x20^3$,... Here we note the use of the 18 and the 20 which is present in their solar and ceremonial calendars. They used a symbol for zero which was used as both a place holder and quantity. Here is an example of how the Mayas would write 4,326.

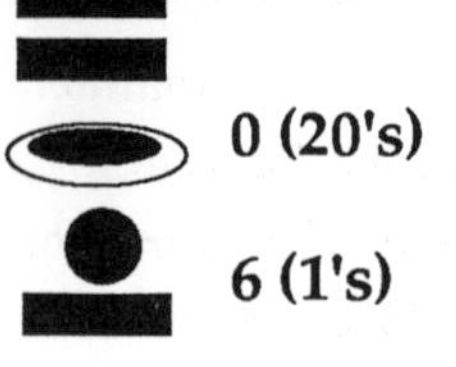

12 (360's)

0 (20's)

6 (1's)

chirality—handedness

Is the universe right handed, left handed, or ambidextrous? No one knows. But the objects of the universe display chirality —handedness. An object is achiral if its mirror image is identical. For example, a sphere or a rectangle is achiral. Chiral objects are those objects which cannot be superimposed on their mirror images. A screw, a tree and a hand are examples of things that are chiral, i.e. are either right or left handed. Molecules have been discovered that form from the right and from the left. We find right and left spirals in plants, shells, bacteria, the umbilical chord, antlers, bone formation, barks of trees. Some areas have a preponderance of one chiral over the other. For example, there are many more right handed people than left, and the same is true of right handed helices of seashells, plants and bacteria. Yet, amino acids are mainly left handed, and helical neutrinos are only left handed. There are some elements that form both ways depending on conditions that exist or mutations that take place. Scientists are studying chirality on micro and macro levels. They are even analyzing the chirality of the four forces that govern the universe — gravity, electromagnetic force, strong nuclear force, and weak nuclear force. Will they discover that nature creates a balance of mirror symmetric forms? Will chirality unlock more secrets of the universe?

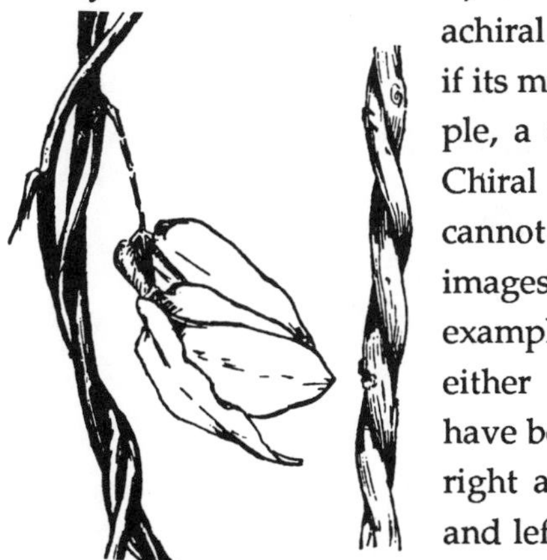

The right handed spiral of wisteria. The left handed spiral of the honeysuckle.

the right and left helical spirals of the antlers of an Alaskan bighorn

the chaos theory—is there order in chaos?

Whether from the oracles and soothsayers of ancient times or from today's increasingly popular tarot cards, astrology charts, fortune tellers and crystal balls, human nature has wanted to know the future. In addition, we seek predictions in such varied areas as:

- weather
- earthquakes
- prices
- stock market
- interest rates
- economics.

Wanting control over our lives, we seek elaborate and costly equipment and methods for prediction. But has our ability for long range predictions of various phenomena been reliable? Do natural phenomena follow a predictable pattern? Do they repeat outcomes in a cyclical fashion? Various sciences for years have relied on the existence of patterns to formulate theories and make general laws. But the *theory of chaos* is shaking up the scientific world. Physicists and other scientists of the traditional schools of thought are beginning to look at the *theory of chaos* more seriously. It may be very distressing to admit that from the very simple events of our world to complicated and major events of the universe the formulas and laws that have evolved over the centuries may not always describe what will take place. Scientists of all disciplines are having to sharpen their mathematical skills and awareness to understand and recognize when *chaos* may present itself in their particular field.

The story of *chaos* begins in the early 1960's with the work of meteorologist and math sophisticate, Edward Lorenz, who set up a very simple experiment using a computer to program various changes in the rising of hot air. What he essentially discovered was that simple minute differences in the initial data could result in enormous differences in the outcomes, i.e. create chaotic occurrences. In weather forecasting this has come to be called *The Butterfly Effect,* — using the analogy that the fluttering of the wings (small changes) of a butterfly in one part of the world could start small air turbulences that could multiply and result in a full scale hurricane in another part of the world. Technically it has come to be described as *sensitive dependence on initial conditions.* The occurrence of small variations in the general picture of weather predictions are continually taking place and their long range global effect is not predictable since it was not feasible to register all possible changes, regardless of how simple or small. These can be considered small errors in information input which can multiply over and over, and thus produce a chaotic event.

An artist's rendition of the Lorenz attractor.

Lorenz created the first picture of the science of chaos when he graphed the results of his experiment in three dimensions. The result was a 3-D spiral-like curve that never intersected or repeated itself. It came to be known as the *Lorenz attractor.*

Strange attractor is a general term for this shape occurring in the *chaos theory.* It can be graphed in multi-dimensional space. The *strange attractor* changes constantly, endlessly looping but never exactly repeating itself. Lorenz's work was published in 1963 in *The Journal of Atmospheric Sciences.* Unfortunately, at that time, scientists in other fields either did not have access to it or chose to ignore it.

It was not until the 1970's that other mathematicians and scientists began discovering similar results[1], especially when computers were used for modeling information. The studies were from a wide spectrum of fields that seemed to be totally unrelated, yet the results were astonishingly similar—the *strange attractor* surfaced over and over. Here are some of the areas of research where the *chaos theory* was discovered:

1) recorded floodings of the Nile

2) earthquakes

3) fluctuations in cotton prices

4) transition points between smooth flowing fluids and turbulence

5) statistical economics

6) noise disruption of electrical current on telephone lines

7) changes in celestial orbits
   - the orbit of Saturn's smallest moon, Hyperion
   - Pluto's orbit
   - the orbits of several moons of Mars and Jupiter

8) variations in Jupiter's Great Red Spot

9) variations in fluid dynamics
   - variations in the flow of a dripping faucet
   - variations in the flow of water in a water wheel

10) variations in non-linear trigonometric functions

11) ecology
   - population fluctuation in the ant population of Sonoran desert highlands
   - fluctuations of measles outbreaks among school children

- insect infestations in Australia
- population fluctuations of the Canadian Lynx

The *Chaos Theory* —Regardless of how simple or complex a recurring phenomenon is, its occurrences have no predictable order; yet there is a form (the strange attractor) of order (though unpredictable) that exists in chaos. Although the variations follow the pattern of a *strange attractor,* the nature of the strange attractor makes it impossible to predict future outcomes[2]. The one important aspect of this double spiral-type curve was that it never intersected itself. Its path was formed from infinite variations generating this orderly aesthetic curve, which does not repeat or cross. In pursuit of the *chaos theory* an entirely new type of scientific experimentation[3] has evolved where mathematics is the major means of exploration in a laboratory harbored inside a computer. The *chaos theory* will require scientists in all fields to, develop sophisticated mathematical skills, so that they will be able to better recognize the meanings of results. Mathematics has expanded the field of fractals to help describe and explain the *shapeless, asymmetrical,*and *randomness* of the natural environment. Add to this the recent findings in the *chaos theory,* and we find mathematicians are on the threshold of the discovery of order out of chaos.

[1]Mathematician Benoit Mandelbrot was one of the first to notice that the *strange attractors* surfaced in a variety of seemingly unrelated fields. In addition, the *chaos theory* was enhanced by Mandelbrot's work with *fractals* — usually computer generated pictures of a system with infinite variations which may be either *geometric* or *random* and whose detail is not lost when magnified. These geometric objects endlessly repeat themselves on an ever shrinking scale, each reproduction is a miniature version of the original (e. g. the snowflake curve). Fractals have come to be labeled the *geometry of nature* and with the use of present day computers, fractal geometry is used to describe aspects of nature (clouds, ginger root, coast lines) that in the past could not have been described using Euclidean geometric methods.

[2]The Earth could follow a predictable orbit for millions of years and suddenly enter a chaotic mode. Bear in mind that millions of years in the life of the universe is a small amount of time.

[3]According to *Science News*, Jan. 26, 1991 issue, scientists at the Naval Surface Warfare Center have been experimenting with ordering the chaotic motion of a magnetoelastic ribbon by making small adjustments to the steady magnetic field, i.e. they are using the small variations that scientists have found to cause chaos to to control it.

the hexa-hexa flexagon

In a broad sense a flexagon can be considered a type of topological model. It's a figure made from a sheet of paper, and has a varying number of faces that are uncovered by flexing the flexagon. They first appeared in the 1890's as toys and magic devices. The creation of the hexa-hexa flexagon in 1934 by Arthur H. Stone actually rekindled mathematical interest in these intriguing figures. At the time, Stone was a 23 year old English graduate student at Princeton University. In order to make his American sheets of paper fit his English notebook, he had to trim a strip off each sheet. He began fiddling with one of these strips of paper, folding them in various ways; and thus came up with the hexa-hexa flexagon. As a result, he and three friends, Bryant Tuckerman, Richard Feyhman, and John Tukey, studied the properties of hexa-hexa flexagons and developed a complete mathematical theory for them. Since then a number of scientific papers have been written on flexagons.*

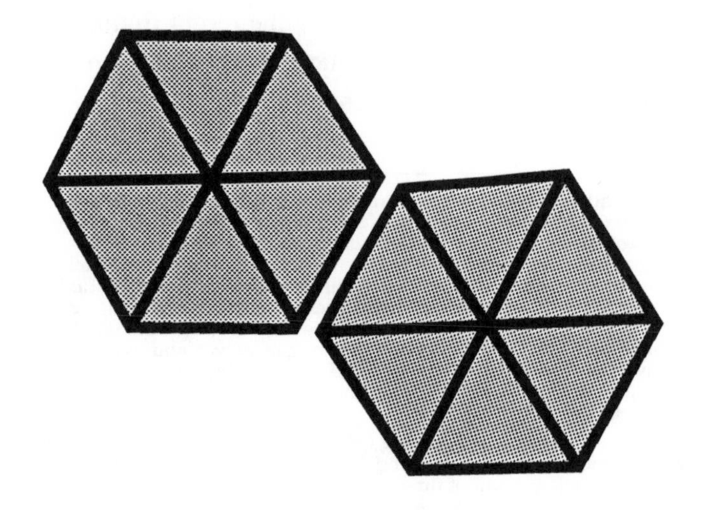

* Martin Gardner's article in *Scientific American,* 1956, O.C. Oakley and R.T. Wisner's article in the March 1957 issue of *The American Mathematical Monthly.*

Front

1

Back

2

3

4

5

fold 2 onto 2

6

7

fold the 3 under so that the 2 is folded onto the 2 on the reverse side

8

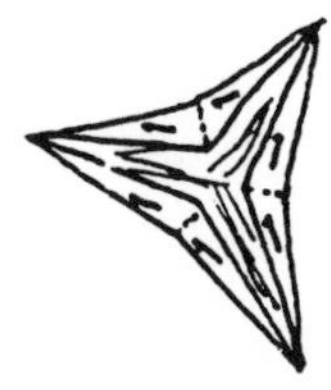

TO FLEX

1) Pinch the flexagon to form the following shape. Open at the center point, just as a flower opens. If the flexagon does not open, pinch along another crease line not used and form the same shape.
2) With practice you will be able to flex it easily.
3) See if you can reveal all six faces by continually flexing.

symmetry—mathematics' balancing act

Place your two hands on a table. Imagine a vertical line passing between your two thumbs; that is a line of symmetry. If a mirror were placed on that line and tilted a bit toward your left hand, it would make a reflection matching the position of your right hand.

The perfect balance one sees and senses in the body of a butterfly, in the shape of a leaf, in the form of the human body, in the perfection of a circle, in the structure of a honeycomb cell may be all attributed to their symmetry.

The concept of symmetry appears in art, nature, the sciences, poetry, and architecture. In fact, it can be found in nearly every facet of our lives. It is something so innate, that we often just take it for granted. Sometimes a form's lack of symmetry may be the quality which makes it appealing. Regardless, when we see some design or sculpture, we (almost immediately) like it or dislike it, and its symmetry, or lack of it probably influenced our feelings.

Mathematical symmetry

From a mathematical point of view, an object is considered to possess line or point symmetry if one can find a line which divides it into two identical parts so that if it were possible to fold along that line both parts would match perfectly over one another.

These are examples of objects with:

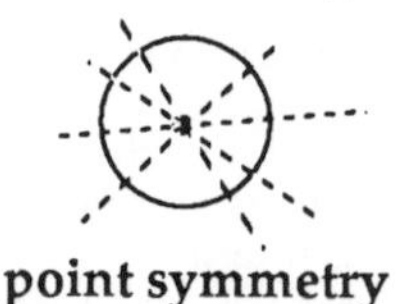

point symmetry

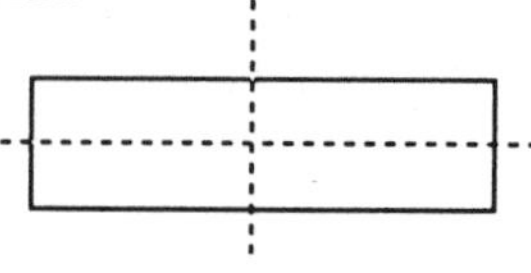

two lines of symmetry

Relections in a pool at Tivoli gardens, Italy.

a line of symmetry **infinite number of lines of symmetry**

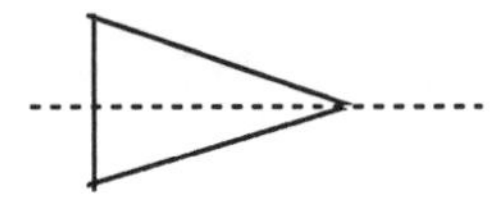

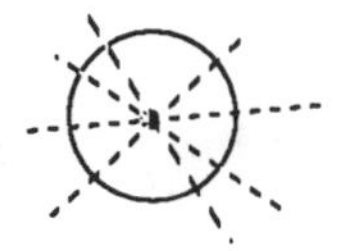

Using algebra, a function's inverse (symmetrical image) can be found by interchanging its x and y coordinates. From these equations, graphs of the function and its inverse can be made, and their line of symmetry will be y=x.

Symmetry and motion

The term, symmetry, is also applied to mathematical relations. For example, the relations, "=" possess symmetry because a=b and b=a both hold true. But the relation "<" does not have symmetry because a<b and b<a cannot both hold true.

One area rarely connected with symmetry is motion. But consider paths of moving objects. If the symmetry of two objects could be established, then one object's moving path could be used to predict another object's motion. Going one step further, consider the actual movement of a four-legged animal. The running position of its legs begin as pictured. When its has reached full speed, its forelegs are parallel as are its hind legs. At this point its movements have line symmetry.

Prior to this, its movement is 180° out of place.

Research has shown that human runners and animals sometimes naturally use a simple form of symmetry. These findings are now being applied in the designs of robots built with legs.

prime numbers & divisibility tests

Discovering prime numbers dates back to ancient times. Eratosthenes (275-194 B.C.) created the prime number sieve for numbers less than a desired amount. Euclid (300 ? B. C.) proved that there is no largest prime number. Today we find mathematicians and computer scientists using computers to establish if a particular number is prime. In August of 1989 computer scientists from the Amdalh Corp's Key Computer Laboratories in Fremont, California established that 391,581x2216,193- 1 is the largest prime find thus far. The previous record holder was 37 digits shorter, a mere 65,050 digits. The algorithms used in determining prime numbers are used in other areas of computer modeling of vast data, such as weather forecasting.

is 5661 divisible by 6?

Divisibility tests are quick tools to answer simple questions about a number's factors. Here are some handy tests to keep in mind:

— How to tell if a number is divisible by —

2— If it is an even number, i. e. ends in 2,4,6,8,or 0.
3— If the sum of the digits of the number is divisible by 3.
4— If the last two digits of the number represents a number that is divisible by 4.
6— If the number meets the tests for 2 and 3.
8— If the last three digits of the number represents a number divisible by 8.
9— If the sum of the digits of the number is divisible by 9.
5— If the number ends in 0 or 5.
10— If the number ends in 0.
11— Find the sum of the digits in even* place- value powers of ten. Find the sum of the digits in odd place value powers of ten. If the difference of these two sum is divisible by 11, the number is divisible by 11.
12— If the numbers meet the tests for 3 and 4.

---

* ... 10^5 10^4 10^3 10^2 10^1 10^0 The even and odd place-value powers refer to the exponents— 0, 1, 2, 3, 4, 5, and so on.

Einstein's "doodles"

$$D = \frac{1}{c}\frac{dl}{dt} = \frac{1}{c}\frac{1}{P}\frac{dP}{dt}$$

$$D^2 = \frac{1}{P^2}\,\frac{P_0 - P}{P} \sim \frac{1}{P^2} \qquad (1a)$$

$$D^2 = \frac{1}{3}\rho\,\frac{P_0 - P}{P_0} \sim \frac{1}{3}\kappa\rho \qquad (2a)$$

$$D^2 \sim 10^{-53}$$

$$\rho \sim 10^{-26}$$

$$P \sim 10^{8}\ L.J.$$

$$t \sim 10^{10}\,(10^{11})\ J$$

It is very special to a see the actual writing and thought process of a genius at work. Einstein's life work dealt with the concepts of light, time, space, energy, matter, space and gravitation and their interrelationships. Here is a replication of the chalkboard work of Einstein from one of his lectures on relativity given at Oxford in 1931.

some patterns of Pascal's triangle

The sum of the numbers down any diagonal to any particular number can be found to be equal to the number directly below in the next row and one place

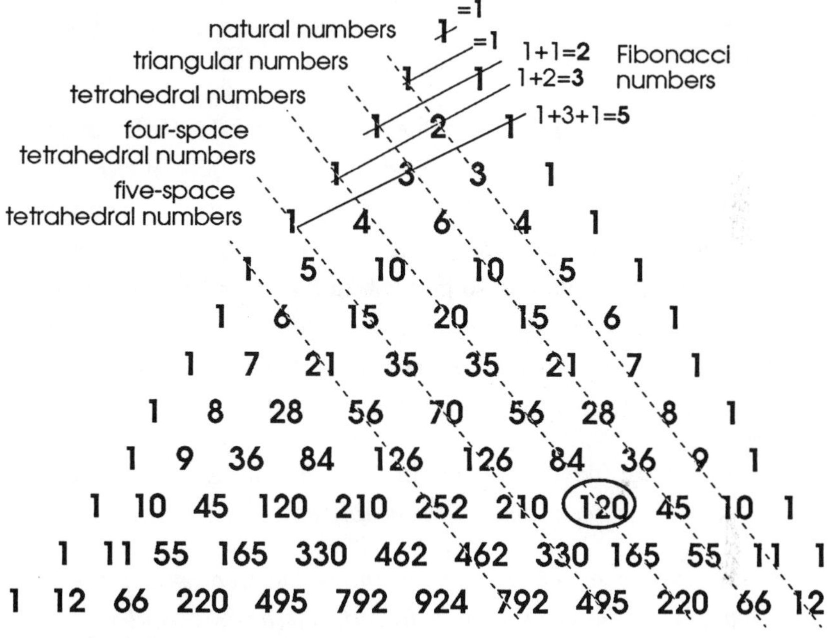

to the left. For example, to sum the triangular number from 1 to 36, all we need to do is look down the triangular diagonal until the number 36 and find the number in the next row below 36 and to the left. In this example the sum is 120.

This method holds true for all the diagonals—natural number diagonal, triangular number diagonal, tetrahedral number diagonal, four space tetrahedral number diagonal and so on.

The Fibonacci numbers also appear in the Pascal triangle*, as illustrated. Apparently Pascal was not aware of this. In fact, this property probably was not noticed until the latter part of the 19th century.

---

* For additional information see pages 242-243.

the pendulum

Be it Edgar Allan Poe's story, *The Pit and the Pendulum,* Umberto Eco's, *Foucault's Pendulum,* the grandfather clock in your house, or the huge pendulum in a natural history museum showing the movement of the earth — all conjure up an undulating, constant almost hypnotic motion.

The story of the pendulum begins in the late 16th century with young Galileo mesmerized by the swinging of the great bronze lamp hanging in the dome of a cathedral. Counting his heart beats and the swing of the lamp to and fro, he noticed the time it took to go back and forth seemed to be the same regardless of the size of the swing. This early observation influenced his study of the properties of the pendulum. In later years he discovered that

1) the period of the pendulum (the swing back and forth) is independent of the amplitude of the swing

2) the period of the pendulum is independent of the mass of the pendulum's bob (thus the bob could be made of wood or lead and it would not affect its period)

3) the period of the pendulum can be determined when the length of the string is known and hence measures time

4) the period can be made any desired length by fixing the length of the string.

In the mid-1600's, the Dutch scientist Christian Huygens researched pendulum motion. He realized the period of the pendulum was not precisely independent of the amplitude of the swing. He knew that minute variations existed. He wanted to refine the motion of the pendulum so that it would

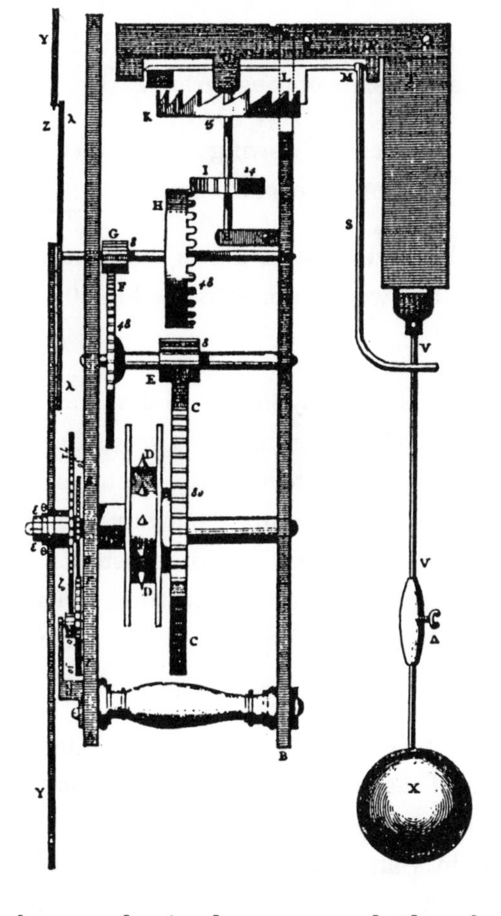

These pendulum illustrations are from Huygen's work of 1673.

The diagram below illustrates the cycloidal jaws that Huygen developed in order to make the pendulum swing in a cycloidal arc.

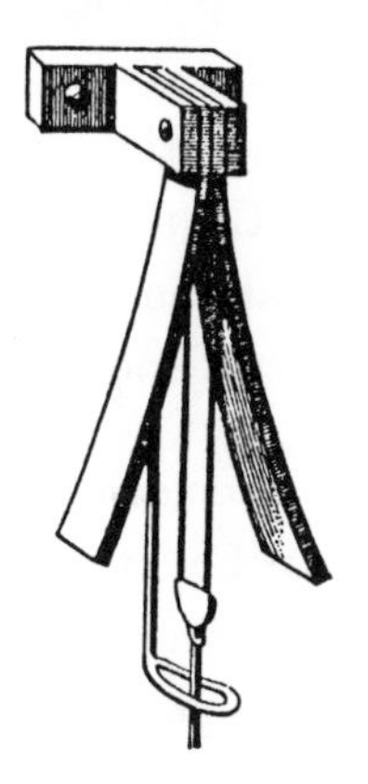

be purely isochronous and therefore be utilized in the development of a clock. He discovered that the solution to the problem lay with the then much studied curve, the cycloid (the cycloid is a curve whose shape is formed by the path of a fixed point on a circle which rolls smoothly on a straight line). Until his discovery, the pendulum motion had been connected to a circular arc. Through his mathematical work, he determined the length of the string that would produce a pendulum whose swing was the path of a cycloid curve. With this information he was able to build the first accurate pendulum clock.

His research also pointed out that different types of pendulums had varied uses. Some of these are:

•*rotary pendulum,* whose swings are actually imperceptible wheel like motions or revolutions

•*torsion pendulum,* whose vertical string is made of wire which becomes a spring when wound up. It is used in 400-day clocks

•*bifilar pendulum,* developed by Lord Kelvin as an earthquake detector. The bifilar pendulum is very effective in measuring variations of the line toward the center of the earth. As a result, the bifilar pendulum demonstrated that the rotation of the earth on its axis is not constant, but slightly increases and decreases from time to time. These variations are caused by the pull of gravity on the earth by the sun and the moon.

•*Foucault pendulum,* developed by the French scientist Jean Foucault in 1851. These are the huge pendulums that are often found in natural history museums (such as the one in Golden Gate Park in San Francisco). The pendulum demonstrates the rotation of the earth, and this rotation makes the pendulum's swing appear to change direction while all along it is the earth that is changing direction while moving along its orbit.

Experiment with the pendulum:
1) vary the height of the swing and see if the period changes.
2) vary the length of the string and see how the period changes
3) see if you can make an isochronous pendulum by testing different lengths of string
4) verify that the mass of the bob does not effect the period

The triple Möbius strip

Ever since Augustus Möbius presented his model of a single-edged single sided object in 1848, mathematicians and math enthusiasts have been "playing" with the Möbius strip. Their discoveries are as fascinating as the original strip. Both of the models below produce the same results.

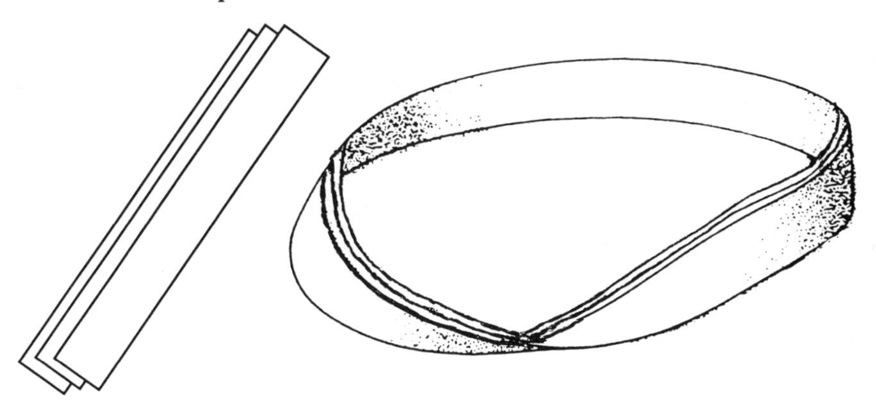

The model above is made from three strips of paper of equal width. The middle strip is of a contrasting color. To make the model give the three strips a half-twist simultaneously, and tape their layered ends together in order. When this model is dislodged compare the results with the model below.

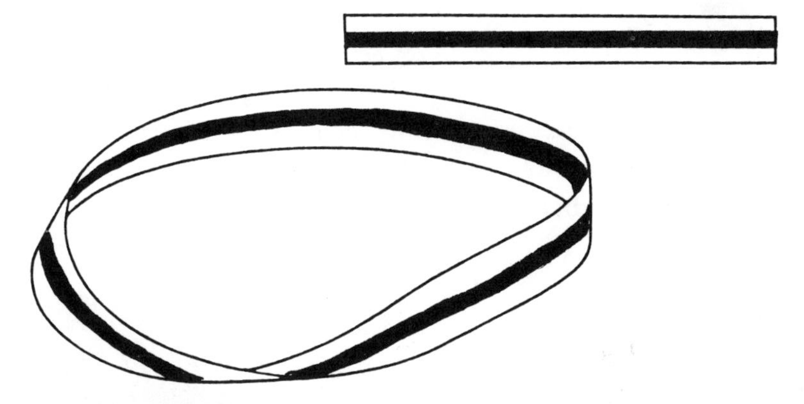

This model is formed from one wide strip of paper which had a thick band colored down its center on both sides of the paper. The band should be one-third the width of the strip. Now form this into a Möbius strip by giving it a half-twist and taping the ends together. With a pair of scissors cut along the end of the color band. Compare your result with the other model.

mathematical treasures from the sea

It is said that the ocean is the source of life. Life forms in the sea, even as on the earth, illustrate a treasury of mathematical ideas.

Look at the vast variety of spirals one finds in the forms of shells. The chambered nautilus and the

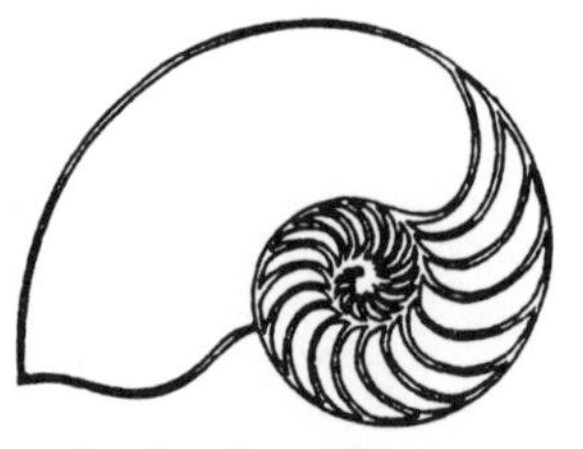

chambered nautillus

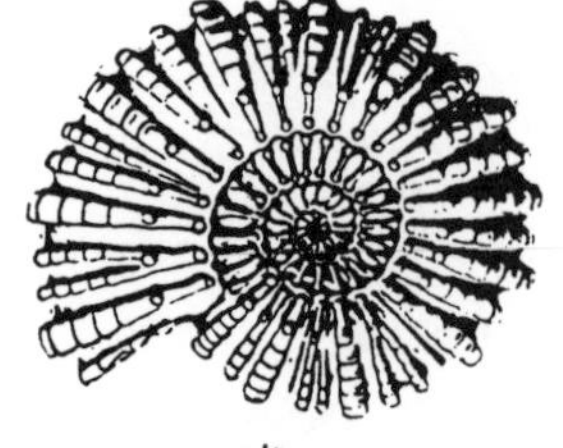

ammonite

wentletrap

ammonite give us the *equiangular spiral.*

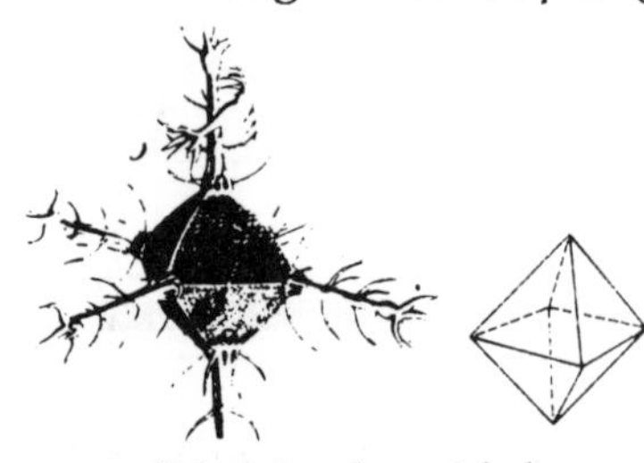

a radiolaria and a octahedorn

The wentletrap and cone shells give us examples of *three dimensional spirals.*

Symmetry abounds in the ocean — *line symmetry* is seen in the clam shell, in body shapes of Paleozoic trilobites, lobsters, fish and many more; the *point symmetry* in the radiolaria and the sea urchin.

Geometric shapes are equally abundant. We see the *pentagon* in the sand dollar and various sided *regular polygons* outlined by the tips of starfish. The *sphere* shape is outlined by the sea urchin.

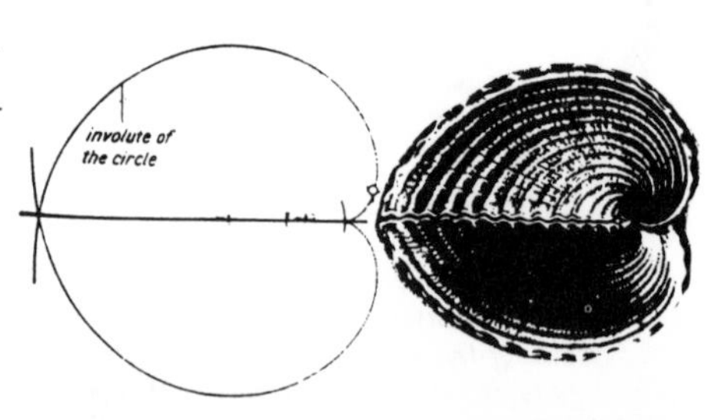

involute of a circle & a cockel shell

The *involute* of a circle is a curve form which resembles the cockle shell. *Polyhedron* shapes are evident in various radiolaria. Rocks have acquired *elliptical* and *circular shapes* as the sea tumbled them in the surf. Random meanders or almost *fractal shaped curves* are formed by coral and the free-form shapes of jellyfish.

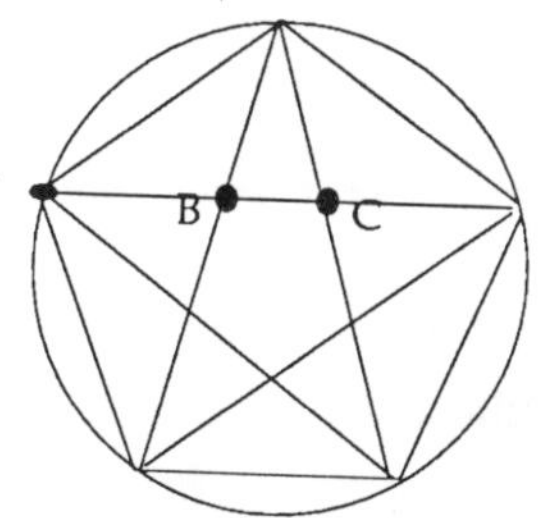

The regular pentagon creates the golden ratio when the pentagram is inscribed. Groups of points (e.g. A, B, & C) form the golden ratio (|AC|/|AB|)=(|AB|/|BC|)

The *golden rectangle* and the *golden ratio* are ever present in sealife. Wherever the regular pentagon is present we also find the golden ratio, as in the intricate pentagon design of the sand dollar. And the golden rectangle is immediately evident in the chambered nautilus and other shell life.

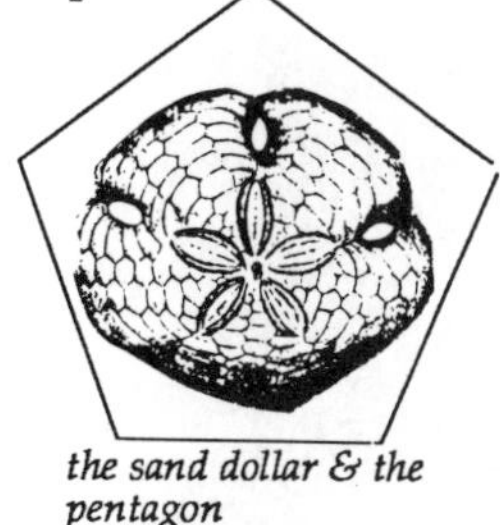
the sand dollar & the pentagon

Swimming underwater in the sea gives one the real feeling of the sensation of three dimensions. One can swim almost effortlessly in the three directions of space.

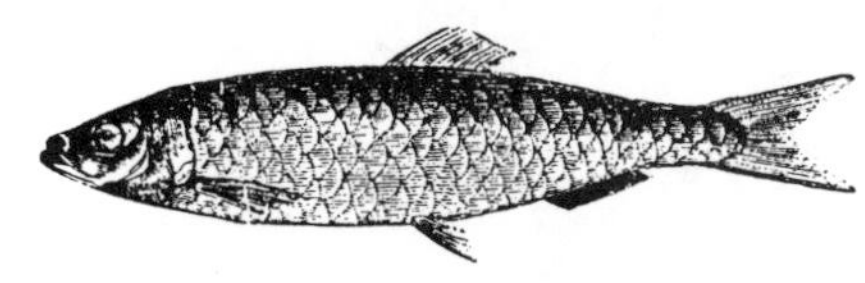
tessellation pattern of scales on fish

We even find *tessellations* in the sea. The numerous patterns of scales of fish are perfect examples of tessellations.

The ocean wave itself is formed from the *cycloid* and *sinusoidal curves*. The wave action resembles perpetual motion. The ocean waves assume so many shapes and sizes—sometimes overpowering and other times gentle but always beautiful, and yet they are governed by mathematical principles —*the cycloid, the sinusoidal curves*, and *statistics*. And finally, was it not perhaps the sands of the sea that stimulated the ancients to formulate the idea of infinity? Each of these mathematical ideas in itself is complex and involved when studied in depth, and yet each gains a new meaning and relevance when discovered in nature.

mathematical knots

At first glance one would think there was nothing special about knots, other than keeping things secured, such as shoes or rigging on a boat. But in in the world of mathematics there exists an entire field called knot theory, and some new discoveries are being made that link this theory directly to the physical world.

Knots have been used by artists and illustrators for centuries. The above is a part of an intricate knot design created by Leonardo da Vinci .

Knot theory is a very recent field of topology. It can possibly link its origin to the 19th century and Lord Kelvin's idea that atoms were knotted vortices that existed in ether, which was believed to be an invisible fluid that filled space. He felt he could classify these knots and arrange a periodic table of chemical elements. Although his theory did not prove true, the mathematical study of knots is a very hot topic today.

What distinguishes mathematical knots from the everyday knots one forms, is that they have no ends. They are a closed type of loop, which cannot be formed into a circle. What mathematicians have been trying to do is to classify knots and be able to distinguish different knots. Some of the important ideas that have been formulated thus far are:

- a knot cannot exist in more than three dimensions
- the simplest possible knot is the trefoil knot, having 3 crossings. It comes in a left and right handed version which are mirror images of one another

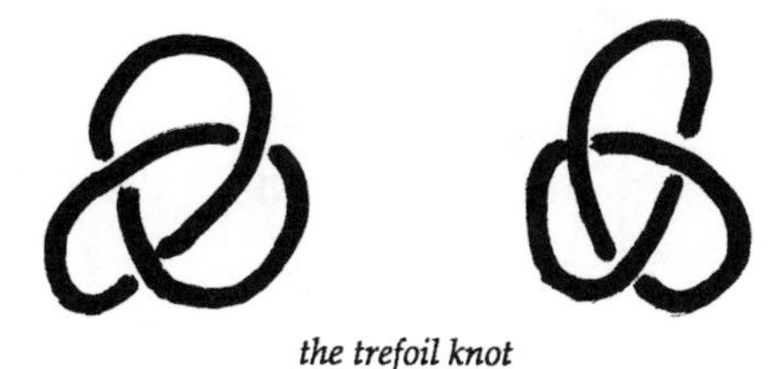

the trefoil knot

- there exists only one knot with 4 crossings

From left to right: the first knot has 4 crossings, the second has 5, the third has 6, the forth and fifth have 7.

- there are only two types of knots with 5 crossings
- over 12,000 knots have thus far been identified with 13 or less crossings, not counting mirror images
- Consider the diagram. These knots are mirror opposites of each other. You would think that since they are opposites, that they would undo each other when brought together. Try it! (They simply pass through one another and remain unchanged.)

- Now look at the Chefalo knot or false knot.

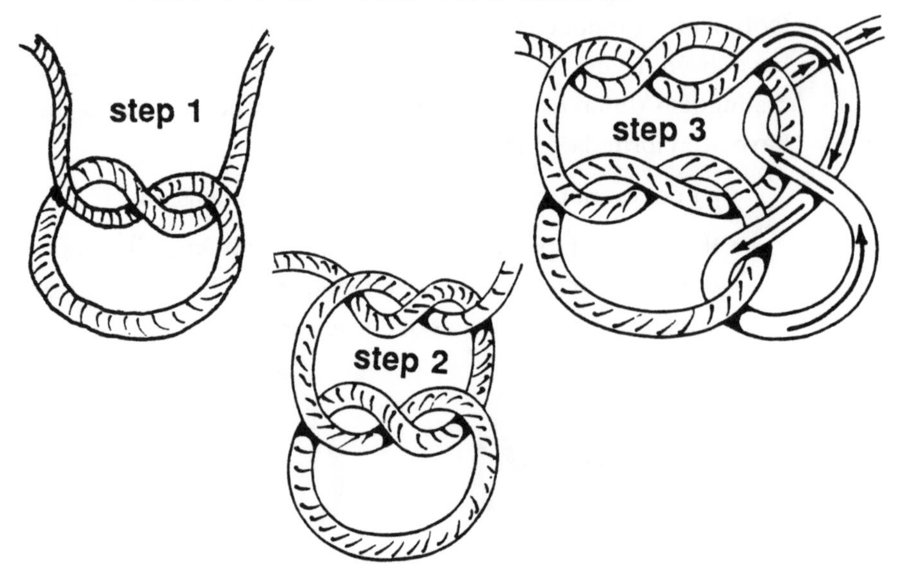

What happens when the ends are pulled? (The knot falls apart.)

Originally, actual models of knots were created in order to study them. To test if two knots were equivalent, one of the models had to be manipulated by hand and an attempt was made to transform it to the other's shape. If it was transformed they were identified as equivalent, if not, no conclusion was possible.

The study of knots in the field of topology tries to explain these various properties. Computers have even come into the picture. The *Geometry Supercomputer Project* [1] is using advanced computer technology to study and produce visual three dimensional representations of mathematical forms and equations, such as torus knots and fractals.

Mathematicians have developed other methods for classifying and testing knots.[2] They now look at the two-dimensional shadow it casts and write an equation describing it.[3] Many recent inroads have been made and some exciting connections discovered linking knot theory to work in molecular biology and physics. Scientists in these fields have used the findings of mathematicians and applied the new techniques in knot theory to their study of DNA configuration. Researchers have discovered that DNA strands can form loops which sometimes are knotted. Now scientists can use findings from knot theory to decide if a DNA strand they are viewing has appeared before in another knot form. They can also determine a sequence of steps in which DNA strands can be transformed to produce a particular configuration and to predict unobserved configurations of DNA. All of these findings may prove very helpful in genetic engineering. Similarly, in physics knot theory proves very helpful when studying the interaction of particles that resemble knots. A knot configuration can be used to describe different interactions that can take place. These are just the beginning discoveries and applications of an emerging mathematical field—knot theory.

---

[1] An international group of mathematicians and scientists who work on the same supercomputer via telecommunications network with pure mathematics to solve challenging problems of geometry. *Science News*, vol. 133, p.12, January 2, 1988 issue.

[2] A "knot" that can be transformed to no twists or crossings i. e. into a loop, is a circle or unknot.

[3] The first such equations were done by John Alexander in 1928. In the 1980's Vaughan Jones made additional discoveries on the equations for knots. *Science News*, vol. 133, p.329, May 21, 1988 issue.

the flexatube

Many mathematical ideas are discovered accidentally, be it when searching for a solution to an entirely different problem, working with a puzzle, a game or even fiddling with a piece of paper. And so it was with Arthur Stone, the creator of the hexa-hexa flexagon (see page 44). In the course of studying flexagons, Stone invented some fascinating objects, one of which was the tetra flexatube. It appears as a flat shaped flexagon, which opens into a tube. The dotted lines indicate creases. He discovered the tube could be turned inside out by flexing along the creases.

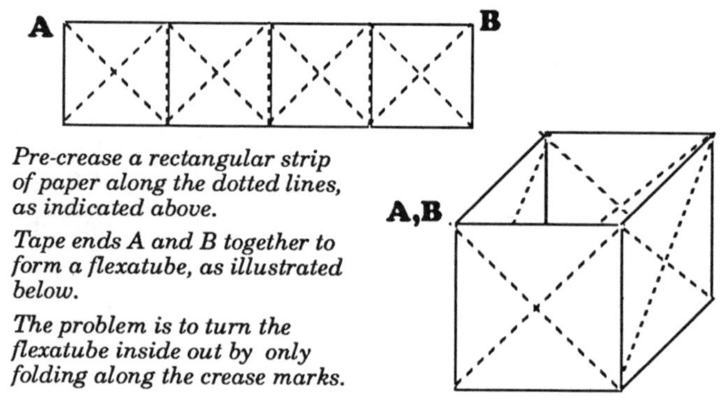

Pre-crease a rectangular strip of paper along the dotted lines, as indicated above.

Tape ends A and B together to form a flexatube, as illustrated below.

The problem is to turn the flexatube inside out by only folding along the crease marks.

Some other types of flexagons that have since been discovered are tri-tetra flexagon (three faced four sided), tetra-tetra flexagon, hexa-tetra flexagon.

Flexagons originated as recreational novelties. Yet their fascinating properties prompted mathematicians to devote time and energy to discover new flexagons, the flexatube, new characteristics, and properties. Although all their applications and uses have yet to be discovered or invented, they still provide an interesting and intriguing manipulative and mental exercise.

Benjamin Franklin's magic line

"In my younger days, having once some leisure (which I still think I might have employed more usefully), I had amused myself in making magic squares."

—Benjamin Franklin

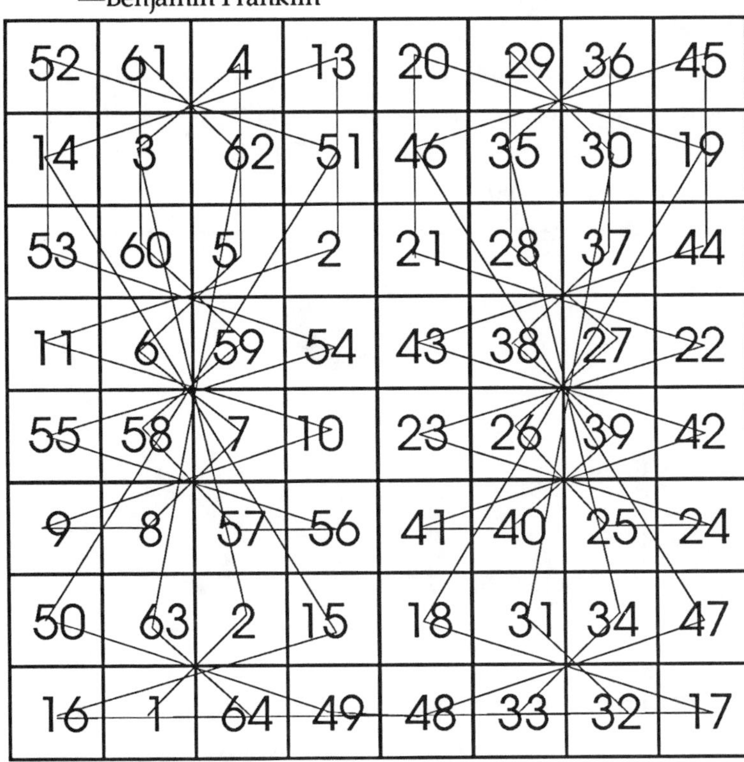

In the first part of the 20th century architect Claude F. Bragdon discovered how magic squares could be used to form artistically pleasing patterns. By consecutively connecting the numbers in a magic square, a symmetrical design was created —called a *magic line.** The diagram illustrates the magic line that is formed by Benjamin's Franklin's magic square.

---

*A magic line is not a line by definition, but rather a series of segments whose endpoints connect the numbers of a magic square sequentially, and form a symmetrical design

where '0' and 'zero' originated

We often take symbols and words for granted. Mathematics has a tremendous number of terms and symbols which have taken hundreds of years to evolve to the forms to which we have grown accustomed. The word *zero* and its symbol *0* provide good examples of this evolution. The actual evolution of the concept of zero is another story in itself, but we can discuss briefly the historical development of its word and symbol.

The symbol *0* first appeared in Hindu writings around 870 A.D. The *0* has come to have many interpretations—the *number* zero, the *origin* on the number line, a *placeholder* in our positional number system, the *identity element* for addition. Originally, in Hindu the word *sunya* meant *empty* or *void,* and in the 9th century A. D. it was used as a place holder in their positional number system. The Arabs translated this word into Arabic, *as-sifr.* In the 13th century, the Arabic word *sifr* was introduced into Germany by Nemorarius as *cifra.* The word was later transliterated into Latin as *zephirum.* In Italian the word became *zeuero,* which closely resembles the word *zero.*

the astrolabe

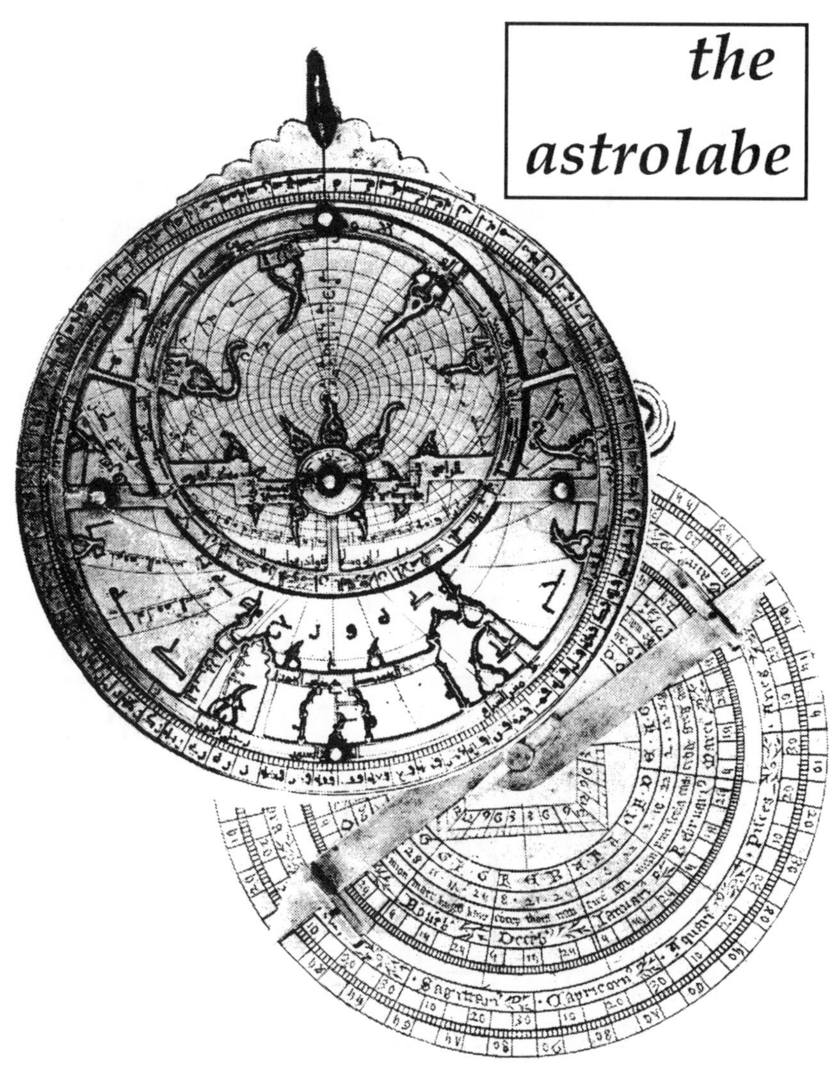

The astrolabe—the pocket watch and slide rule of the world.

The astrolabe, originally a Greek invention, was improved upon by the Islamic people. It measured the elevation of the sun or of another star. They would use this information in conjunction with star maps, and would make additional computations to determine information about the sunrise, sunset, latitude, the time for prayer, and the direction of Mecca for the Islamic traveler.

the eight checkers puzzle

This old puzzle has been adapted by commercial puzzle manufacturers over the years. There are over four million ways that eight pieces can be placed on a 64 square unit checkerboard.

The 8 Checkers Puzzle

Place eight checkers on a checkerboard so that no two checkers lie in the same row, column or diagonal.

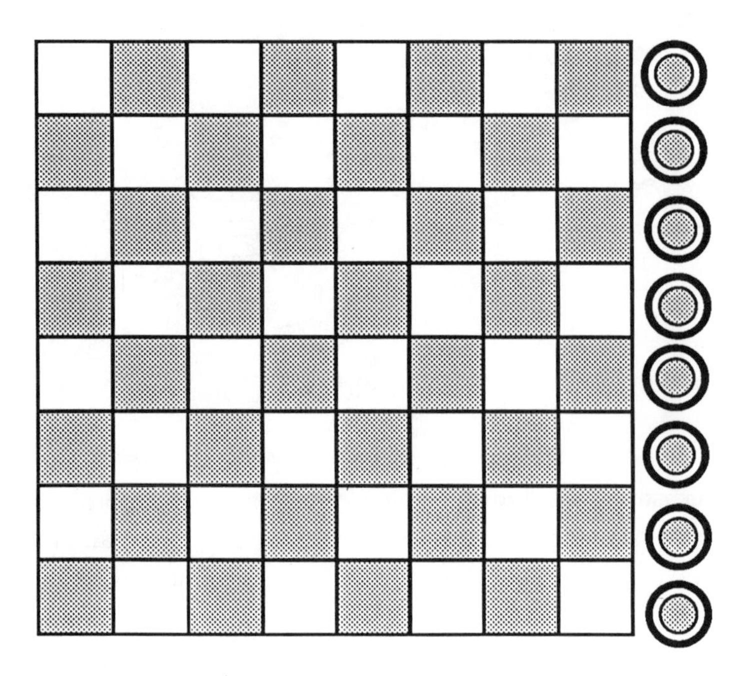

For a solution, see the appendix.

the game of squayles

The wonderful thing about stick games is that they can be played anywhere. One doesn't need a special board or special pieces. But the game of squayles does require a keen eye. To begin squayles 31 sticks are arranged as shown in the diagram.

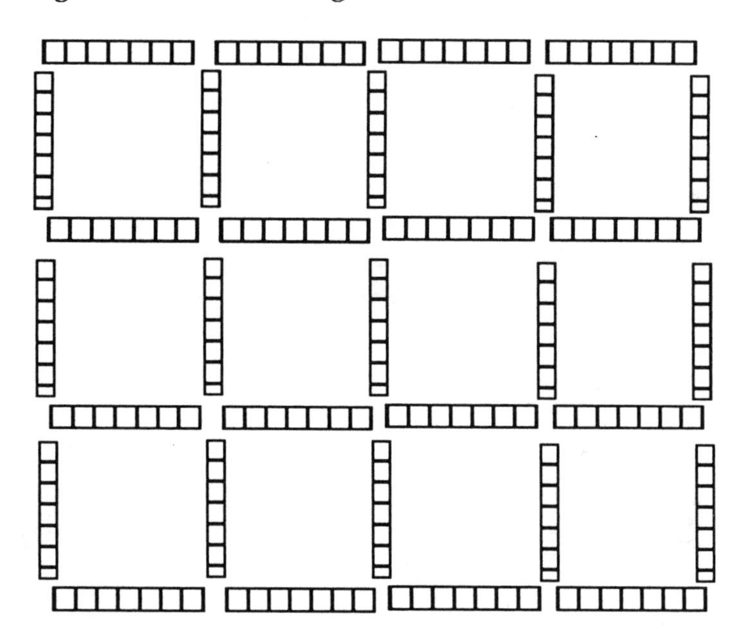

Taking turns, each player may take as many sticks as he or she wants as long as they are adjacent to one another.

For example—

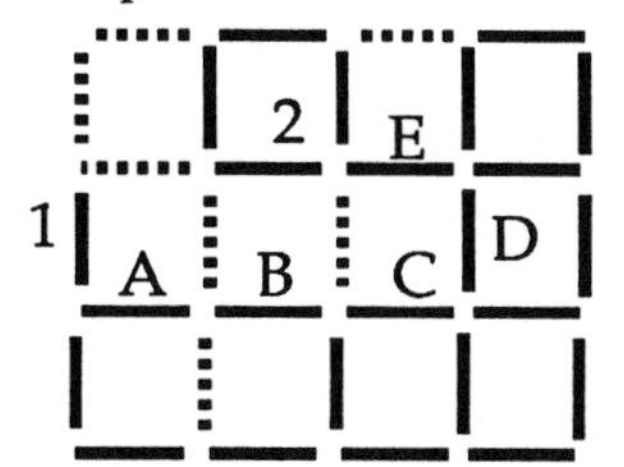

Since sticks 1 and 2 are not adjacent they cannot be taken directly. But you could get sticks 1 and 2 by first sequentially taking 1, then A, B, C, D, E, and then 2.

The winner is the player taking the last stick.

ENJOY!

the circle helps Queen Dido

From the Roman poet Virgil, we learn the story of Queen Dido. She was the daughter of a Tyrian king, who had to flee to Africa after her brother murdered her husband. There she begged for some land from King Iarbus. Suspicious of her request, King Iarbus asked how much land she desired. She replied by requesting as much land as the hide of an ox would enclose. Since this seemed a minor request, the King granted her wish.

Being a shrewd woman, she cut the hide into very thin strips and decided the greatest area would be enclosed if she formed a circle. It was thus she founded the city of Byrsa (oxide), later known as Carthage.

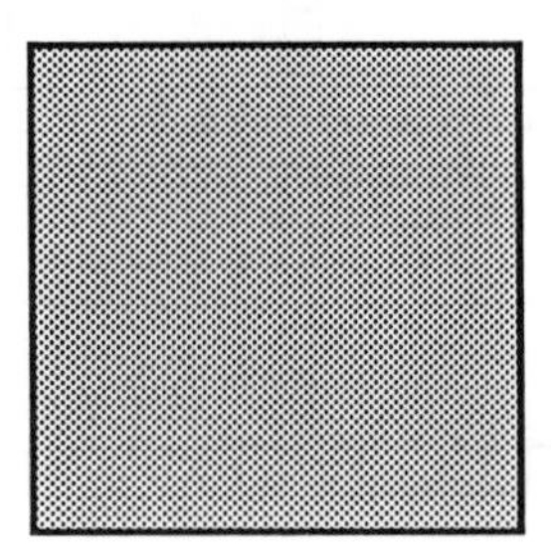

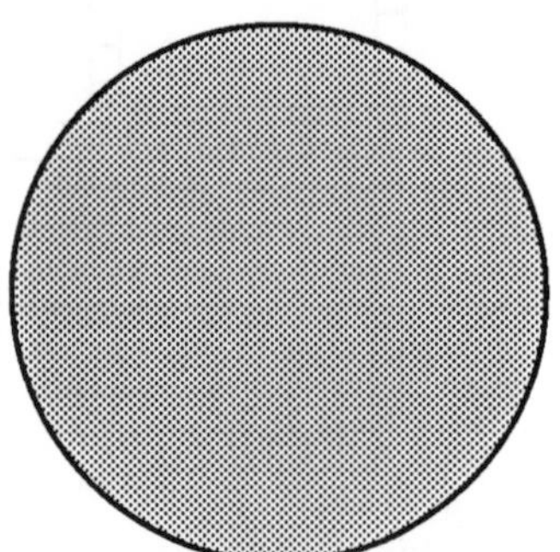

A square of perimeter equal to the circumference of a circle, encloses less area than the circle.

Proof: *Assume the perimeter of the square is x. Then the length of a side of the square is $x/4$ and its area is $(x/4)^2=x^2/16$.*

A circle with circumference x, has diameter equal to x/π. Thus its radius is $x/2\pi$, and its area would be $x^2/4\pi$.

Therefore, the area of the square, $x^2/16 <$ the area of the circle, $x^2/4\pi$. (note: $16>4\pi$)

The quadratrix curve was a mathematical discovery that resulted from the pursuit to solve the three construction problems of antiquity (squaring a circle, trisecting an angle, and duplicating a cube).

quadratrix— the curve that both trisects an angle and squares a circle

Hippias discovered the quadratrix around 420 B. C., and found that it could be used to both trisect an angle and square a circle.

The quadratrix is formed in the following way—
A square is constructed in which its side, AB, rotates from its base position in the counterclockwise direction. As AB rotates at a constant angular velocity, another segment (with endpoints on sides AD and CB) moves parallel to AB at a constant linear velocity. The points of intersection of these two moving segments form the points of the quadratrix curve. The following ratios are always equal— $(m<XAB/m<DAB)= (|XX'|\backslash|DA|)$.

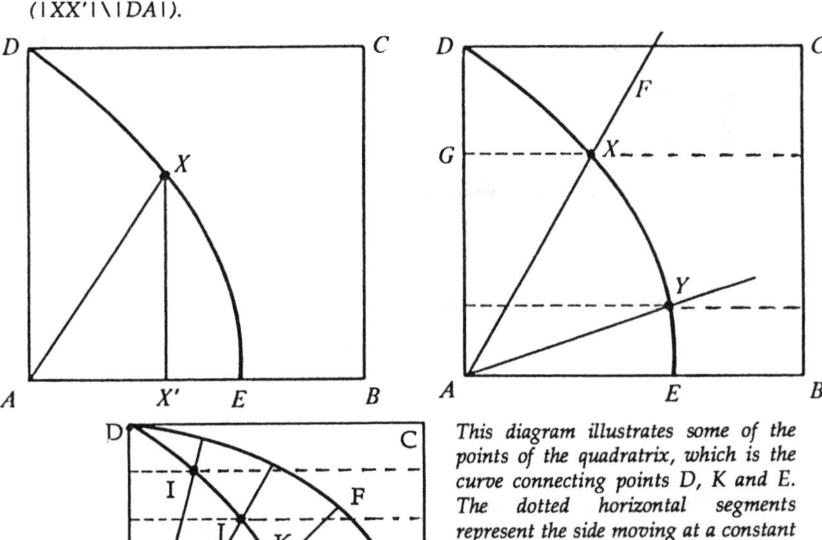

This diagram illustrates some of the points of the quadratrix, which is the curve connecting points D, K and E. The dotted horizontal segments represent the side moving at a constant linear velocity, and the radii along the circular arc DFB represent the segment moving at a constant angular velocity. Their points of intersection are D, I, J, K, L, M, E—points of the quadratrix.

Lewis Carroll's window puzzle

Mathematician Charles Dodgson is better known as Lewis Carroll, author of **Alice's Adventures in Wonderland** and **Through the looking Glass.** Here is one of the many puzzles he developed.

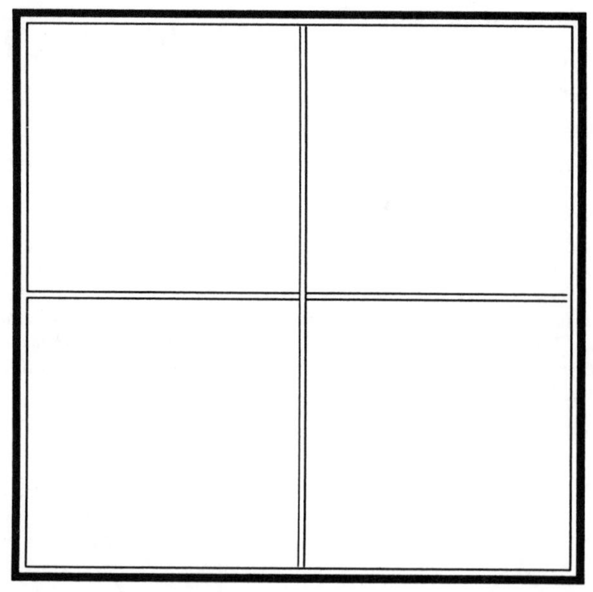

This window gave more light than what was desired. How can it be changed so that it will remain a square but give half the amount of light? Curtains, drapes or any coverings cannot be used. Its height and width must remain 3′.

For a solution, see the appendix.

fractal time

We often think of inanimate objects as immobile objects. Yet just as the wind, ocean waves, rivers are definitely moving so do rocks, glass, plastic. In fact, any material or substance is moving. We may not see or measure the immediate movement because the movement is taking place at a molecular level. We may also not have associated the deterioration of an object's composition as movement — but its molecules are moving or realigning themselves — especially if they are under different stresses (e.g. pressure, temperature, electrical or magnetic fields). Today in industry , the study of changes in manufactured goods, such as plastic, glass, rubber, silk, and other goods, is very important, since it plays a role in the shelf life of the material.

In crystalline material, change proceeds at an exponential rate, similar to the half-life in the decay of radioactive substances. The molecules of non-crystalline materials (amorphous material) change or move over a span of varying times. Some take seconds while others take years. This reorganization phenomenon of non crystalline material could be described by the term *fractal time*. *Fractal time* is based on the same idea as that of a fractal. The magnification of a geometric fractal reveals a miniature version of the larger shape. Consider *time* in the form of this type replication. The time intervals of a material's molecule reorganization differ and occur similarly to the stages of a fractal's duplication process. The time analogously depends on the stage at which the material is being viewed. Thus, mathematics of fractals will play a vital role in the study of materials' changing process, and industry will use the findings to improve a product's shelf life.

codes & ciphers

The study of cryptology spans centuries. One of the earliest examples dates back to ancient Sparta, when Spartans would write secret messages using a scytale (stick) and a strip of leather or parchment. The strip would be wound tightly around the scytale in a spiral fashion. A message would then be written on the strip. Unwound the strip appeared to have an irrelevant sequence of letters. It could be reconstructed again if wrapped around a scytale of the exact size and shape.

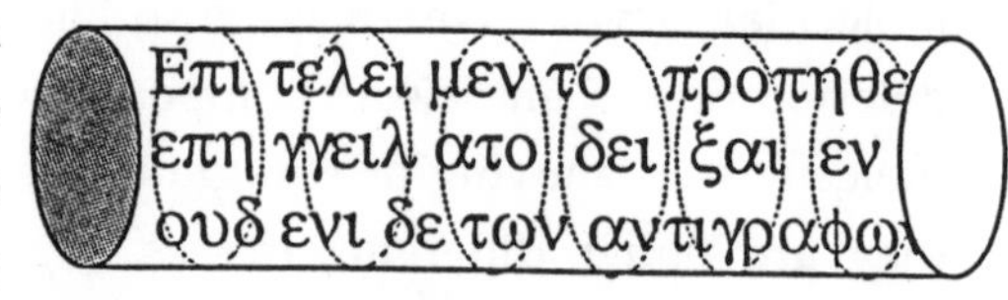

The centuries that followed are rich with examples and stories of people creating codes, ciphers and cipher devices and other people decoding and deciphering the secret messages. Julius Caesar would sometimes write secret messages by substituting each letter of a message with one a fixed number of letters down the alphabet. The Islamic people began to develop organized approaches to deciphering. European countries during the Renaissance established special groups of people to work on deciphering secrets from other countries. Some manuals and treatises were written during the 14th and 15th centuries, but the most advance work in cryptanalysis has been done during the 20th century. Here we find the work of William and Elizebeth Friedman, two very famous cryptologists. *

**Both William and Elizebeth Friedman's original careers were far afield from cracking codes and ciphers. William Friedman was a graduate in genetics, while Elizebeth Smith (married name Friedman) graduated with a degree in English. They were persuaded by Col. George Fabyan to join his Riverbank laboratories in 1915 to do projects in their particular areas of expertise. A special project, which was to try to prove Francis Bacon was the true author of Shakespeare's work, led them to begin their work in cryptology. In the winter of 1916-1917, they studied intensely every bit of information that they could find on cryptography. Through Elizebeth's work on decoding radio messages, the U.S. Department of Justice was able to prosecute major narcotic rings in the 1930's. William did invaluable work for the U.S. government during both World Wars. His work and books (some remain classified information) expanded the field of cryptanalysis beyond the mere counting of frequency of letters and discovered and applied revolutionary statistical techniques to cracking codes and ciphers.*

Since mechanics and the field of cryptography are closely related through the development of cipher devices, along with creating codes and ciphers goes the building of cipher devices to mechanically encipher or decipher messages. These machines are made to systematically restructure messages. Various devices played a major role in World War II. For example, the Japanese *Purple machine* was virtually reconstructed from scratch (no information or part of the device was available) by William Friedman's team, thus providing invaluable information to the Allies. Other notable accomplishments during this period were— the breaking of the so called impregnable German military code called *Enigma,* by England's mathematician Alan Turing — the cracking of the Japanese fleet code by the U.S. Navy. Coming up to present, newspaper headlines during the Iran-Contra scandal brought to public light a high security device. This present day cipher device is used to make top secret phone calls secure by somehow scrambling the messages over the wave lengths. The best selling novel, *The Name of the Rose,* which deals with a cryptic code in its plot , was made into a film.

From Rudyard Kipling's story, ***The First Letter,*** *in his book,* ***Just So Stories.*** *In Kipling's words—"The letters round the tusk are magic—Runic magic—and if you can read them you will find out something new." These letters actually form a substitution cipher.*

Codes, ciphers and cryptograms are in themselves a breed of logic problems. There are various methods and approaches on how to attack certain types of problems. Similarly, understanding the creation of codes and ciphers along with working numerous ones give one an automatic advantage in reaching the solution.

We saw how the Spartans used the scytale. The arrangement of the keys of a typewriter has also been used as a cipher device. The *grille,* a card with holes punched in it, was a code device developed by Cardinal Richelieu in the 17th century. When placed on a seemingly innocent note, the holes would reveal the letters of a secret message. Even Thomas Jefferson developed a cipher device that was cylindrical in shape and composed of 36 wheels, each printed with the alphabet on it and able to be rotated independently from the other wheels. Today we have at our finger tips the most advanced tool to both aid in the breaking and creating of codes and ciphers—the modern computer. The power of this modern day tool is phenomenal when used by an expert.

Understanding how ciphers and codes are created is essential to learning how to break a code or cipher. It requires the same perseverance and ingenuity as solving a very difficult math problem. In essence, it is a special type of math problem. History has shown that solving these problems is definitely not merely recreational. Some are linked to the turning points of wars, while others solve some of the mysteries of medicine and genetics, such as the understanding of the double helix code of a DNA chain of molecules. Lastly the work of cracking codes and ciphers is directly connected to modes of communication—be they deciphering hieroglyphics, creating artificial intelligence, or receiving and decoding messages from outer space.

Types of cipher systems

1) *transposition cipher* keeps the original letters of a message, but changes their position in some systematic fashion. For example, the message— *Meet Martha on Monday in front of the bridge*— can be scrambled by writing each word backwards and making each break three letters long —*tee meh tra mno yad nom nit nor ffo eth egd irb.* But such a transposition cipher is rather easy to decipher.
Try a different type of pattern, for example:
M T R A M D I R
EE MA TH ON ON AY NF ON
T T B D
OF HE RI GE
Now write all the bottom letters first followed by all the top ones and break them into groups of 5 letters each, and add any dummy letters to round off the last word to five letters. So we get—
EEMAT HONON AYNFO NOFHE RIGEM TRAMD IRTTB ***DFGHR***
This cipher now becomes somewhat harder to crack.

Placing your message in the spaces of a square is another method used to create a transposition cipher. Then create a path through the letters.
2) *Substitution cipher:*
In *monoalphabetic* each letter is replaced by only one other letter. For example, if the letters of the alphabet are substituted with the ones following: the keys of a typewriter or any scrambling you choose—
A B C D E F G H I J K L M N O P Q R S T U V W X Y Z
Q W E R T Y U I O P A S D F G H J K L Z X C V B N M
so the message —
MEET MARTHA ON MONDAY
becomes—
DTTZ DQKZIQ GF DGFRQN

Or, using Julius Caesar's method of replacing every letter by the third one down the alphabet we get
PIIW PDUWKD RQ PRQGDB

Using a *key word* is another substitution technique. Suppose we select the word MATH, then—
ABCDEFGHIJKLMNOPQRSTUVWXZ
becomes—
MATHBCDEFGIJKNOPQRSUWXYZ

and the message—
MEET MARTHA ON MONDAY is
KBBS KMQSEM NL KNLHMY

polyalphabetic substitution uses more than one alphabet to encipher a message, and different symbols can represent the same letter and the same symbols can represent different letters. This system was first described in print by Blaise de Vigenère who was a French cryptographer of 14th century. *Example*— Choose a key word, for example *MATH*. Now proceed as follows:
key word is repeated—
MATHM ATHMA THMAT HMATH

*MEETM ARTHA ONMON DAY**GF***
message lined under key word (*G&F are dummy letters*).

Code letters are found from the table
*YEXAY AKATA HOYOG KMY**ZM***

	ABCDEFGHIJKLMNOPQRSTUVWXYZ
A	ABCDEFGHIJKLMNOPQRSTUVWXYZ
B	BCDEFGHIJKLMNOPQRSTUVWXYZA
C	CDEFGHIJKLMNOPQRSTUVWXYZAB
D	DEFGHIJKLMNOPQRSTUVWXYZABC
E	EFGHIJKLMNOPQRSTUVWXYZABCD
F	FGHIJKLMNOPQRSTUVWXYZABCDE
G	GHIJKLMNOPQRSTUVWXYZABCDEF
H	HIJKLMNOPQRSTUVWXYZABCDEFG
I	IJKLMNOPQRSTUVWXYZABCDEFGH
J	JKLMNOPQRSTUVWXYZABCDEFGHI
K	KLMNOPQRSTUVWXYZABCDEFGHIJ
L	LMNOPQRSTUVWXYZABCDEFGHIJK
M	MNOPQRSTUVWXYZABCDEFGHIJKL
N	NOPQRSTUVWXYZABCDEFGHIJKLM
O	OPQRSTUVWXYZABCDEFGHIJKLMN
P	PQRSTUVWXYZABCDEFGHIJKLMNO
Q	QRSTUVWXYZABCDEFGHIJKLMNOP
R	RSTUVWXYZABCDEFGHIJKLMNOPQ
S	STUVWXYZABCDEFGHIJKLMNOPQR
T	TUVWXYZABCDEFGHIJKLMNOPQRS
U	UVWXYZABCDEFGHIJKLMNOPQRST
V	VWXYZABCDEFGHIJKLMNOPQRSTU
W	WXYZABCDEFGHIJKLMNOPQRSTUV
X	XYZABCDEFGHIJKLMNOPQRSTUVW
Y	YZABCDEFGHIJKLMNOPQRSTUVWX
Z	ZABCDEFGHIJKLMNOPQRSTUVWXY

Locating the code letters is almost like graphing points on an x,y plane. The *second E* from the word *MEET* corresponds to the letter *T* in the word *MATH* above it. Therefore find **T** on the horizontal alphabet (x-axis) and **E** in the vertical alphabet (y-axis), and where these two intersect is the letter *X* in the table.

Even more complicated are *digraphic substitutions* that deal with pairs of letters. Some cryptographic systems even use syllables, words phrases, sentences or perhaps even paragraphs. They also use code books of a special vocabulary.

3) *combination of transposition & substitution:*
This method can be very complicated, but not impossible to crack, if enough of a message is given.

space telescope —math error sends Hubble's aim off by trillions of miles

On April 25, 1990, the Hubble Space Telescope was launched from the space shuttle Discovery. In essence, NASA placed an observatory above the Earth's atmosphere. Although there are many great telescopes at various observatories on the Earth, they are limited by interference from the Earth's atmosphere. The Hubble Space Telescope operates in the vacuum of space. From its vantage point it can see many times further into space and with increased clarity. The Hubble Space Telescope weighs approximately 25,500 pounds, is 43 feet long, 14 feet in diameter, and can observe objects at wavelengths that are largely invisible from the Earth. It was named after astronomer Edwin P. Hubble (1889-1953), who was responsible for finding the first direct evidence of the expanding universe and with deriving a formula with which astronomers can estimate an object's distance from the Earth by measuring its speed.

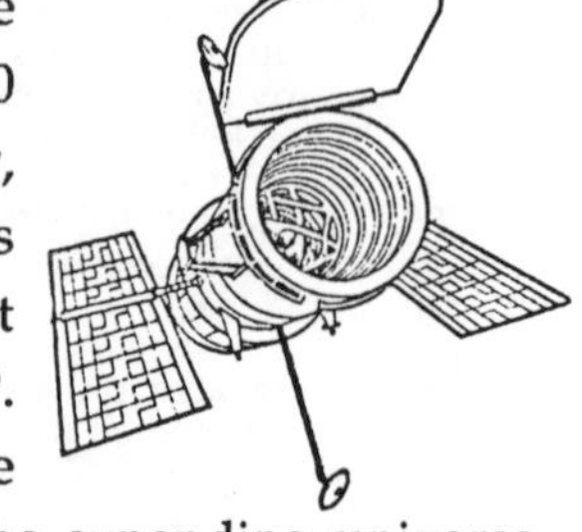

Just as early sailors plotted their course by the stars, the Hubble Space Telescope relies on certain guide stars to orient its observations. Unfortunately, its first space pictures were delayed due to a simple math error. When astronomers designed the pointing instructions for the telescope, they relied on star charts that were made in 1950. Naturally, during the past 40 years the stars have moved from the Earth's vantage point. Realizing this mistake, scientists adjusted their instructions for this change, but instead of subtracting the amount, they added—a simple check book balancing error. In astronomical terms, the half-degree* error threw the telescope's aim off by trillions of miles.

---

*Astronomers went from 18 to 36 arc minutes, rather than from 18 arc minutes back to 0. An arc minute is one-sixtieth of a degree.

mathematics of a forest fire

Beginning as a branch of probability in the late 1960's, *interacting particle system* is an expanding mathematical frontier. Mathematical models and computer simulations are used to study the spread of various natural occurrences. Mathematicians model a collection of wandering or randomly scattered particles—such as trees—represented on a checkerboard grid. Each marked cell or cluster at a grid's center represents trees. The cells are either burnt, burning or untouched. Each increment of time and burning cell has the probability of spreading to one of its four neighboring cells unless the trees in that cell are already burned.

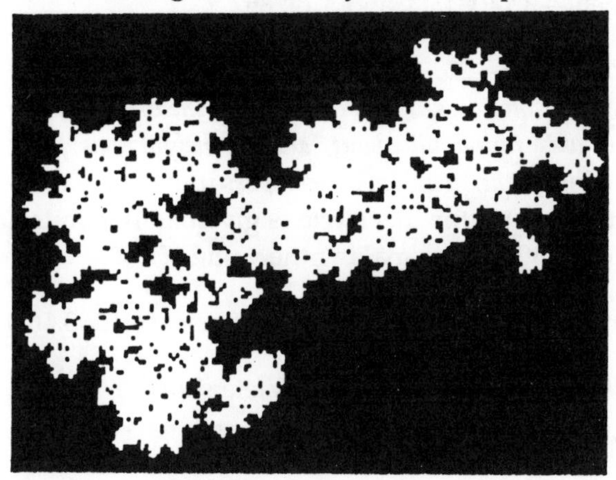

computer generated illustration of areas of burned forest and spread of fires

As of now, these models are not as complicated as real life situations. Similar models are also being applied to the spread of an infectious disease — in which case each cell represents an individual that is healthy, ill, or immune.

Mathematicians are studying different degrees of probability and how computer simulations progress under such probability. As more complex information can be structured into mathematical models, results and predictions from such work may play an essential role in the understanding and manipulation of certain natural phenomena.

early estimates & expressions for π

In 1700 B.C., the Egyptians value for π, 256/81=3.16050..., came from their area formula for a circle, namely— A=(d-(1/9)d)2.

Archimedes (287-212 B.C.) proved that π was between 3 10/71 and 3 1/7, namely—3.140845... and 3.142857...

In the Bible, Kings I, vii—23,

> *"And he made a molten sea, ten cubits from the one brim to the other: it was round all about, and its height was five cubits: and a line of thirty cubits did compass it round about."*

we see that the diameter for this circle is given as 10 cubits and its circumference is 30 cubits, making the value of π=3.

In ancient China, Ch'ang Hong (125 A.D.) gave π as $\sqrt{10}$=3.162...

In 1592, French mathematician François Vieta showed

$$\pi = 2\left(\frac{1}{\sqrt{\frac{1}{2}}\cdot\sqrt{\frac{1}{2}+\frac{1}{2}\sqrt{\frac{1}{2}}}\cdot\sqrt{\frac{1}{2}+\frac{1}{2}\sqrt{\frac{1}{2}+\frac{1}{2}\sqrt{\frac{1}{2}}}}\cdots}\right)$$

In 1655, English mathematician John Wallis showed that

$$\pi = 4\left(\frac{2\bullet4\bullet4\bullet6\bullet6\bullet8\bullet8\bullet10\bullet10\bullet12\bullet12\ldots}{3\bullet3\bullet5\bullet5\bullet7\bullet7\bullet9\bullet9\bullet11\bullet11\bullet13\ldots}\right)$$

The German mathematician Gottfried Leibniz (1646-1716) proved

$$\pi = 4\left(1-\frac{1}{3}+\frac{1}{5}-\frac{1}{7}+\frac{1}{9}-\frac{1}{11}+\frac{1}{13}-\frac{1}{15}+\frac{1}{17}-\frac{1}{19}+\ldots\right)$$

In 1873, English mathematician William Shanks published the first book evaluating π. In it he carried π out to 707 places. Since computers were not available, one can imagine how tedious the work was. It was not until 1948, that John W. French Jr. of United States and D. F. Ferguson of Great Britain published π to 808 decimals places. Their estimation revealed that Shanks' value had its first error at the 528th place.

The modern computer revolutionized estimating π, and the π goes on and on.

Pythagorean Triplets

Is there a formula that will generate Pythagorean triplets*? The ancient Greeks dealt with this question. Here's what they discovered.

If m is an odd natural number, then
$((m^2+1)/2)^2 = ((m^2-1)/2)^2 + m^2$ *will give a Pythagorean triplet.*

a, b & c — $a^2+b^2=c^2$
5, 12, & 13, so that $5^2+12^2=13^2$

For example, if m=17 replacing this value for m, the formula yields $145^2=144^2+17^2$. Thus the Pythagorean triplet is 17, 144 and 145.

This formula was known to the Pythagoreans, but another form of it was derived by Plato in which any natural number could be used for m.

Plato's formula *was* —

$$(m^2+1)^2 = (m^2-1)^2 + (2m)^2, \text{ where } m \text{ is a natural number.}$$

Are there any triplets that are not given by this formula?

(Try 7, 24, 25. Since m^2+1 and m^2-1 differ by 2, the triplet 7, 24 and 25 cannot be given by Plato's formula because 24 and 25 differ by 1.)

Euclid's method for finding Pythagorean triplets was—

If x and y are integers and if $a=x^2-y^2$, $b=2xy$, $c=x^2+y^2$, *then a,b, and c are integers such that* $a^2+b^2=c^2$

These formulas should generate all possible Pythagorean triplets!

*A Pythagorean triplet is a set of three numbers that satisfy the equation that the sum of the squares of two of these numbers equals the square of the third number.

one step beyond the Pythagorean theorem

Although the Pythagorean theorem has been around for thousands of years, the varied proofs and ideas that spin-off from it continue to fascinate. There are various ways in which the theorem can be stated.

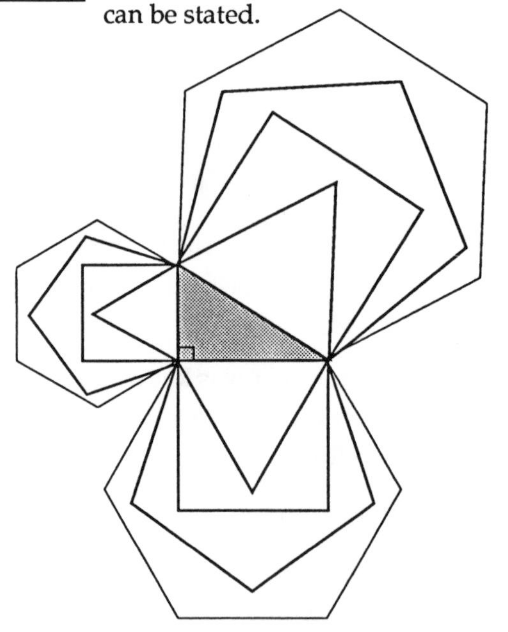

For example—
Given a right triangle, the sum of the squares of the legs will equal the square of the hypotenuse

or— *Given a right triangle, the sum of the areas of the two squares on the legs equals the area of the square on the hypotenuse.*

Now let's discover a variation of the last form. Suppose instead of summing the areas of squares on the legs and on the hypotenuse, we sum the areas of other similar geometric figures, and determine if the theorem still holds true. Here's how it would work for semicircles. Will it always work for similar objects?

area of circle A is $(1/2)\,(a/2)^2\pi=(a^2\pi)/4$

area of circle B is $(1/2)\,(b/2)^2\pi=(b^2\pi)/4$

area of circle Cis $(1/2)\,(c/2)^2\pi=(c^2\pi)/4$

totaling their areas: $(a^2+b^2)\pi/4=(c^2)\pi/4$

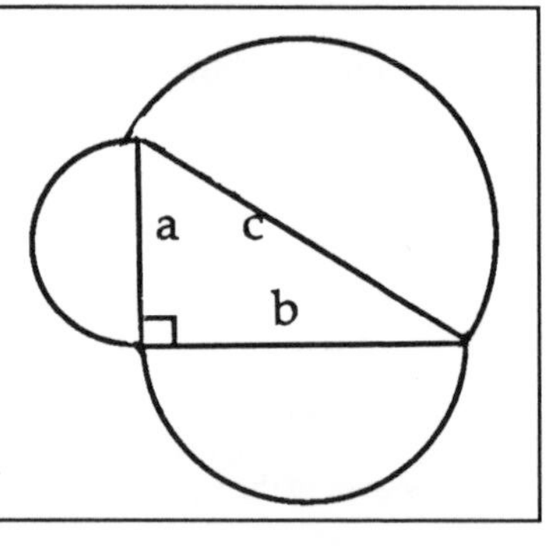

tying polygons

Fiddling with a strip of paper has produced interesting mathematical models.

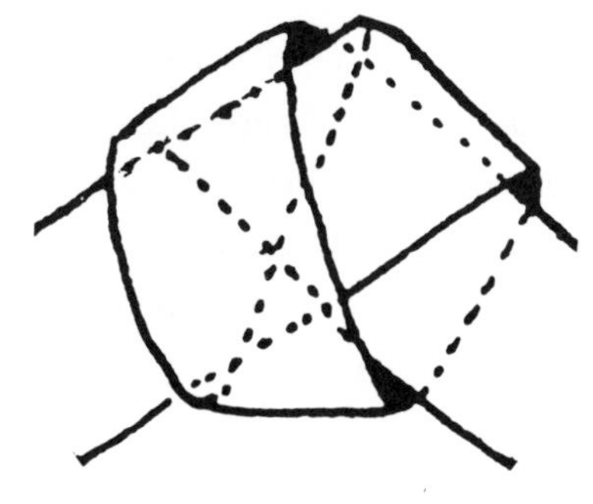

Let's discover how to form a pentagon and a hexagon from a strip of paper. Take a strip of paper and simply tie it as illustrated. Amazingly a regular pentagon is formed.

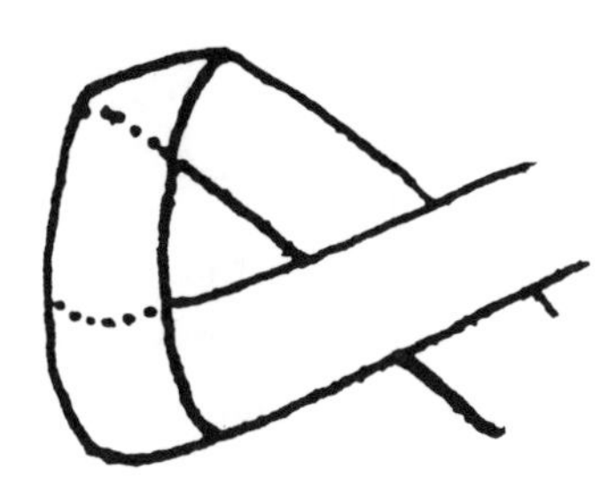

A hexagon is just as easily made. Take two strips of paper and tie a square knot. The illustration shows an easy way to manipulate the strips into a square knot.

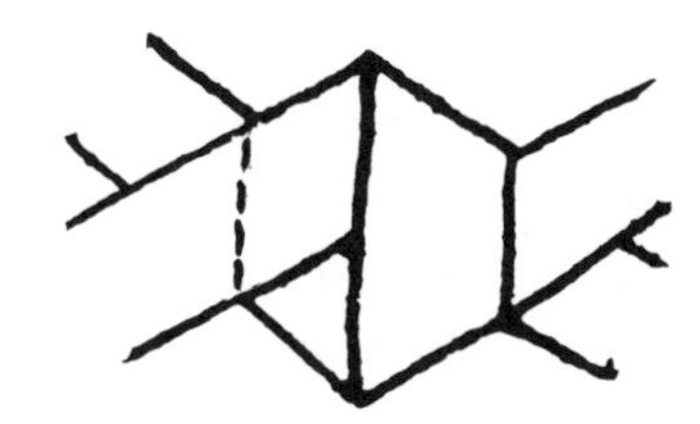

Why not try forming other polygons from strips of paper?

the world of Riemannian geometry

Almost from the inception of Euclid's Fifth Postulate (also called the Parallel Postulate), there were mathematicians who felt it could be proven and therefore was not a postulate. In 1854 G. F. Bernhard Riemann presented a paper on spherical (or elliptical) geometry. Using an indirect proof of the Parallel Postulate[1], he stated the negation of the Parallel Postulate as follows —*through any point there are no lines which can be drawn parallel to a given line.* Riemann also decided to see what would happen if he changed the interpretation of another postulate of Euclid, namely — *straight line may be produced to any length in a straight line.* He changed it to *a line is boundless but not infinite* i. e. it has no ends but is finite in length. Such properties can exist in a geometry on a sphere in which all lines are great circles[2]. Studying a sphere, note that any two great circles always interesect in two points and thus no lines (great circles) are parallel. In spherical geometry we also find that the sum of the angles of a triangle total more than 180° and as a triangle's area increases the sum of the angles increases.

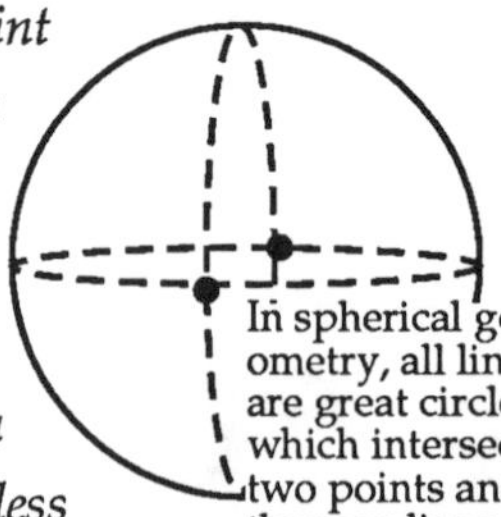

In spherical geometry, all lines are great circles which intersect in two points and thus, no lines are parallel.

Where does such a world exist? Perhaps it is our universe? If the mass of our universe is great enough so that the gravitational pull is able to stop its expansion since the Big Bang, it would eventually contract, shrink, and its form would be depicted in a sphere shape. Over billions of years this spherical universe would eventually shrink to the size of a point of infinite heat and density. If the gravitational pull is not great enough to contract it, then perhaps it would reach a point of balance where expansion would just come to a halt.

[1]One way in which the Parallel Postulate can be stated is—*through any point not on a given line there is only one line parallel to the given line.*

[2]A great circle is a circle on the sphere whose center is the center of the sphere.

twin snowflakes?

Walking in a snow flurry you are in the midst of some wondrous geometric shapes. The snowflake can be one of the most exciting examples of hexagonal symmetry in nature. Yet snow crystals, as they are referred to by researchers, can also be shaped in columns, needles, plates or lumps. The infinite number of combinations of patterns, explains the common belief that no two snowflakes are alike. Nancy C. Knight of the National Center for Atmospheric Research in Boulder, Colorado, may have discovered the first set of identical snowflakes. They were collected on November 1, 1986 on a glass plate that was coated with oil and then exposed in a cloud for 11 seconds at an altitude of about 20,000 feet over Wausau, Wisconsin. The plate was kept cold, until the plane landed and the crystals photographed. They were column-shaped crystals, which are called lacunas.

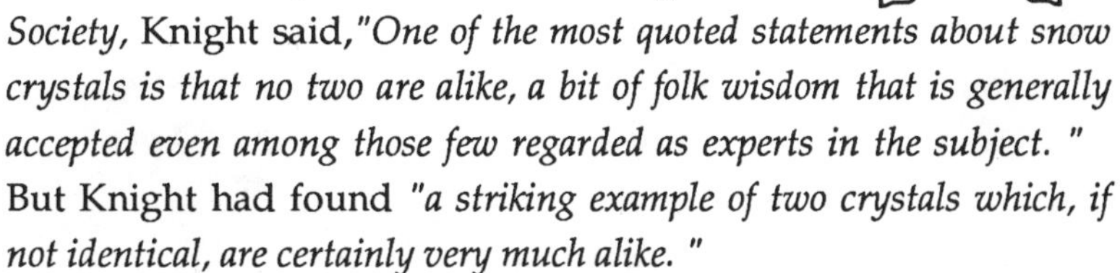

In a letter published in the May, 1988 issue of the *Bulletin of American Meteorological Society*, Knight said,"*One of the most quoted statements about snow crystals is that no two are alike, a bit of folk wisdom that is generally accepted even among those few regarded as experts in the subject.*" But Knight had found "*a striking example of two crystals which, if not identical, are certainly very much alike.*"

She continued, "*in many years of snow-crystals collection the author has seen no other example of such crystals, nor are any given in the standard references.*"

Is it possible to calculate the probability of this phenomenon occuring?

computers & art

Over the centuries, history has shown that artists and their works have been influenced by the knowledge and use of mathematics. We find the intentional use of the golden rectangle and the golden ratio in ancient Greek art, especially by the famous sculptor Phidias. The mathematical concepts of proportions and ratios, similarity, perspective, projective geometry, optical illusions, symmetries, geometric shapes, designs and patterns, limits and infinity, and now computer science have affected various facets and periods of art from primitive to modern.

There are some works of art that might not have been created if it were not for the artist's knowledge and use of mathematics. For example, the tessellations created by Moslem artists and the extension of these geometric forms to include animate objects by M.C. Escher would not have been realized if the artists had not pursued their study and discovery of properties of tessellations including concepts of congruences, symmetries, reflections, rotations, translations of geometric forms. Escher formulated for himself the laws of periodic space-filling, but his first attempts at the art of tessellation were failures because he had not yet discovered the necessary mathematics.

Artists such as Albrecht Dürer, at times resorted to mechanical devices based on the mathematics of projective geometry to create some of their works. Today artists are exploring a new form or medium of art that is linked to mathematics—the computer. Until recently computer art was produced by mathematicians, scientists,

perspectograph

engineers, and just about everyone but artists. Initially, we were flooded with a great deal of art that resembled curve stitching, line work and optical illusions.

With today's advanced computers and software, Leonardo da Vinci might have rendered this sketch on a computer.

Today, computers play a major role in commercial art. A computer skilled artist with advanced software can transform graphic art ideas for advertisements to multiple variations by changing typestyles, introducing various colors, scaling sizes of objects, rotating or flipping objects, duplicating various parts of an object in minutes. All of these changes would have taken the graphic artist hours if not days in the past.

Engineers, architects and other designers have not hesitated to embrace the computer in their creations. With just a few clicks of a mouse a building could easily be modified, a plane could be rotated to show all possible angles, cross-sections could be added, parts added or removed effortlessly. In the past this work was slow and painstaking.

Artists over the centuries have sought different mediums to create their works—watercolor, oils, acrylic, chalk, etc. There are some artists who feel the computer is an artificial means that lacks freedom of spontaneous expression. They prefer the direct contact of their hands with the medium of their choice, rather than working by means of electricity on a keyboard and screen or with stylus and electronic tablet and screen. Others view the computer as a challenge.

With improved software and hardware, colors can be mixed on the screen. Rigid computer lines can be easily softened with any shaped curve. Changes can be made from an oil painting effect to a watercolor, and brush shapes can be changed at whim. Minute areas can be magnified and reworked more easily. Parts can be erased or cut out and pasted elsewhere in the work. The artist is definitely in charge of his or her creative work. The result can be either printed or entered on video, paper or film. Film leaves the option of enlarging it to any size. Perhaps a printer will be designed capable of capturing the texture of the work the artists create. Or perhaps we should consider this a new form of texture in itself.

Esteemed artists have displayed computer art in well known international galleries, but it is still labeled *computer art.* Such works have yet to be simply labeled *art.*

What would Leonardo da Vinci think of the use of computers in art? Given his fondness for innovation*, we can assume he would not have snubbed the use of the computer. He said "...no human inquiry can be called science unless it pursues its path through mathematical exposition and demonstration." His works reflect that he extended this to his art– for example, the predominant use of the golden rectangle in many of his works and the concepts of projective geometry in his masterpiece, *The Last Supper.* One form of art should not be considered better than another, just different. Artists should be free to choose any means or medium.

*His notes and various innovations were used by artists to enhance and facilitate their artwork. Leonardo's mathematical inclinations led him to invent various types of special compasses capable of producing parabolas, ellipses and proportional figures. He is also credited with the invention of the perspectograph, used by artists (such as Albrecht Dürer) to help draw objects in perspective.

how Archimedes trisected an angle

The ancient challenge to *trisect an angle using only a compass and straightedge* led to the development of some fascinating mathematical ideas and instruments. One such was Archimedes' sliding linkage.

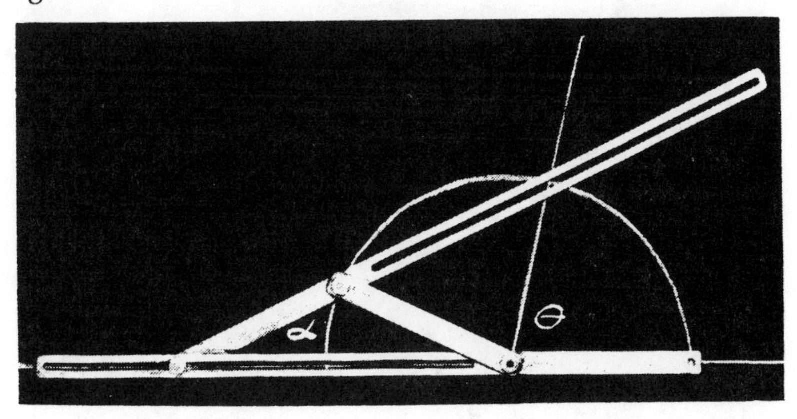

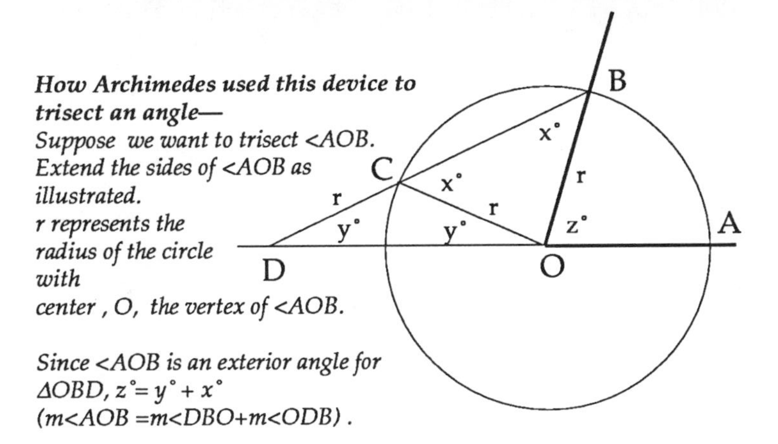

How Archimedes used this device to trisect an angle—

Suppose we want to trisect <AOB. Extend the sides of <AOB as illustrated. r represents the radius of the circle with center , O, the vertex of <AOB.

Since <AOB is an exterior angle for ΔOBD, $z° = y° + x°$ *(m<AOB =m<DBO+m<ODB) .*

Similarly, <BCO is an exterior angle for ΔCDO, thus $x° = y° + y° = 2y°$ *(m<BCO=m<CDO+m<COD)*
By substitution, we get $z° = y° + 2y°$*, so that* $z° = 3y°$*.*

Therefore, $y°$ *is 1/3 the size of <AOB, and thus <AOB trisected.*

fractal clouds

Nature never ceases to amaze with wonders that lend themselves to fractal replication. This photograph of the sun burning the edges of clouds looks like a fractal computer replication.

Fractals are considered fractional dimensions. In Euclidean geometry, a point is zero-dimensional, a line is 1-dimensional, and a plane is 2-dimensional. What about a jagged line? In the field of fractals, a jagged line's dimension is considered to lie between 1 and 2. Starting with a line segment and subdividing it into 3 sections as is done with a snowflake curve, its dimension will lie between 1 and 2. If we start with a rectangle, a 2-dimensional object, and subdivide it into 4 sections and construct a pyramid over its middle section, a fractal of dimension between 2 and 3 is formed.

fractals & ferns

The geometry of fractals has come to be known as the geometry of nature. Fractals provide a mathematical means to describe objects in nature which do not conform to Euclidean objects. Think of a geometric fractal as an endless generating pattern—the pattern continually replicates itself but in a smaller version. Thus, when a portion of a geometric fractal is magnified it looks exactly like the original version. In contrast, when a portion of a circle is magnified it begins to appear less curved.

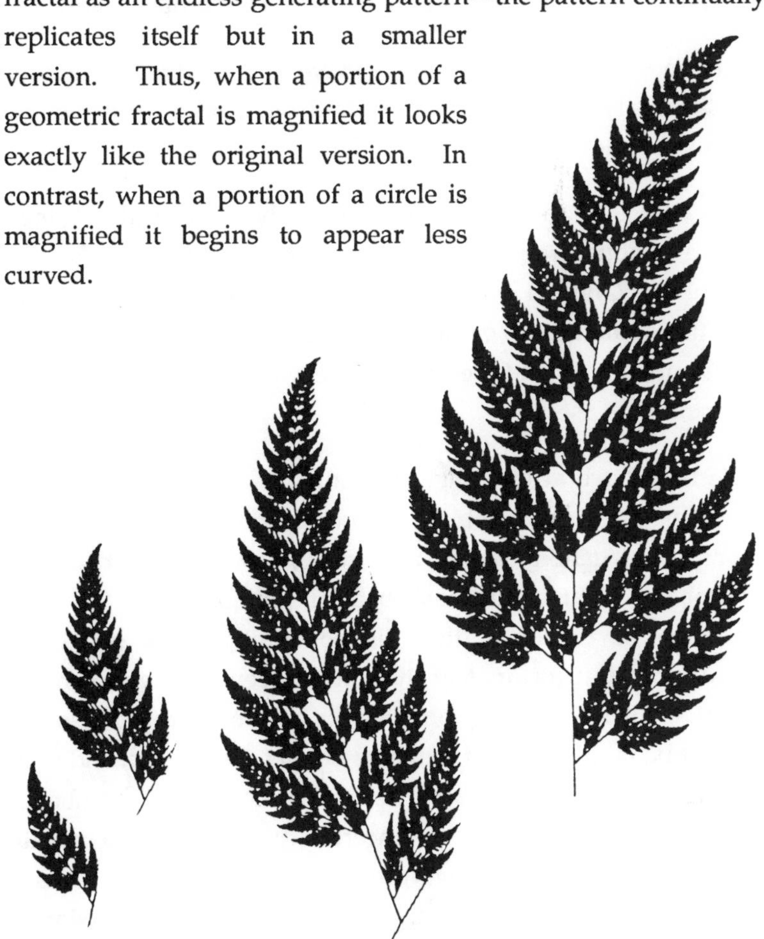

A fern is an ideal example of fractal replication. If you zero in on any portion of the fractal fern, it appears as the original fern leaf. A fractal fern can be created on a computer.

the evolution of numbers

To condense the development of numbers seems almost sacrilegious. Where, when and how did natural numbers, whole numbers, integers, rational numbers, irrational numbers, imaginary numbers, real numbers, complex numbers evolve?

Like most ideas in mathematics they either evolved by accident, by necessity, by curiosity, or by a driving need for exploration into the realm of ideas.

It is difficult to imagine what it must have been like trying to solve various problems and being confined to a specific set of numbers. Granted, often problems are made with set restrictions on the domain and/or range, but at least we are aware of the existence of other types of numbers and such problems just become exercises.

We now have at our finger tips all the complex numbers, but imagine tackling a problem whose solution was the value of x for the equation $x+7=5$ and not knowing about negative numbers. What would have been the various reactions — the problem is defective, there is no solution, the equation is incorrect.[1] But fortunately there have been bold and confident mathematicians willing to take risks and to believe the solution might be in the realm of numbers not yet discovered, and finally going one step beyond and defining a new set of numbers. Imagine the excitement or uncertainty of creating a negative number to solve the above problem. Of equal interest is the testing of these new numbers to see if they obey the axioms of the existing set of numbers.

It is almost impossible to pin down specific times and locations of the origins of different numbers, but we can reconstruct

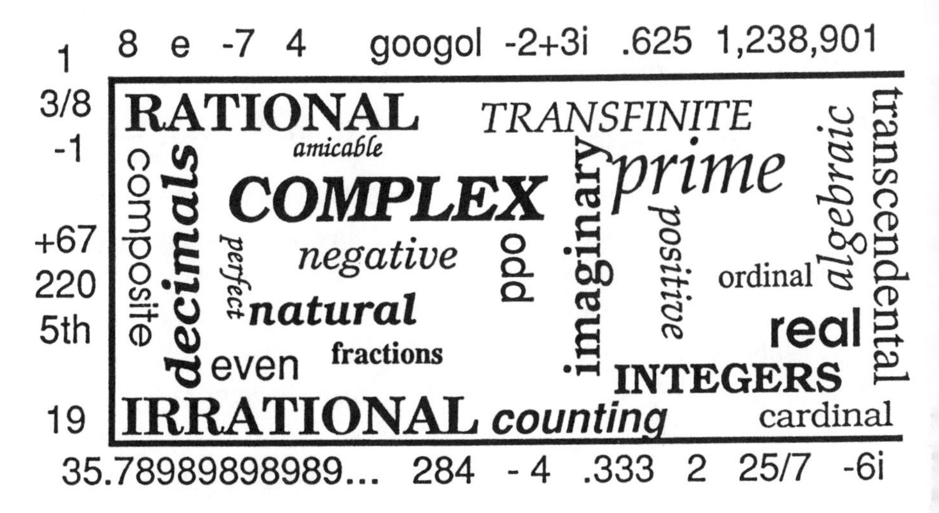

similar type of problems and scenarios that would require the invention of a new type of number.

Peoples in different parts of the world used only the natural numbers for centuries on end. Perhaps they had no other needs. Of course the respective symbols and systems of writing the natural numbers differed from culture to culture.[2]

The appearance of the first zero dates back to the second millennium when zero was represented on a Babylonian clay tablet. It was first a space in the clay tablet and later either of the two symbols, or , were used for zero. But here zero was used as a placeholder, rather than for the number zero. The Maya and Hindu number systems were the first to use their symbols for zero both as a placeholder and as the number zero.

The rational numbers may have been the next to evolve. People had a need to divide whole quantities, such as a loaf of bread. Though symbols may not have been devised to represent these numbers, the ancients were aware that fractional quantities existed. The Egyptians, for example, used the glyph for *mouth* , , to write their fractions, for example

was 1/3, was 1/10, was 1/223.

The Greeks used lengths of line segments to represent various quantities. They knew that all points on a number line could not be taken up by the natural and the rational numbers. At this time we find the introduction of irrational numbers. The question remains —Was the first irrational number $\sqrt{2}$ [3], which is the result of applying the Pythagorean theorem to a right triangle with legs of length 1; or was it phi, $(1+\sqrt{5})/2$, the golden ratio arising from their use of the golden rectangle? Regardless, we know irrational numbers were used at this time.

History reveals that the discovery of new numbers to solve old problems simultaneously creates new problems. The discovery of a set of new numbers was one thing, but deriving its definition and logical system, along with its acceptance and general use always took years to evolve.[4] Negative numbers were difficult for European mathematicians to accept, even into the 17th century. The use of the square root was not restricted to any particular set of numbers, thus imaginary numbers were created with $\sqrt{-1}=i$. Cultures over much of the world had polynomial equations that required the use of imaginary numbers in their solutions. One such equation is $x^2 = -1$. Now to tie all these numbers together a universal set had to be devised. Here enter the complex numbers, which appeared as solutions to such quadratic equations as $x^2+2x+2=0$. Complex numbers (numbers of the form $a+ bi$, where a and b are real and $i=\sqrt{-1}$) were introduced during the 16th century. All the above numbers can be classified as complex, e.g. a real number is a complex number whose imaginary part is 0 and an imaginary number is a complex number whose real part is 0.

The imaginary and complex numbers became more concrete, when they were geometrically described. As the Greeks had begun to describe the real numbers and their location on the number line, the complex numbers were described using a complex number plane. Every point on a plane can be assigned one and only one complex number and vice versa. Thus the five solutions of equation, $x^5=1$, could be graphically represented.

Since the complex numbers describe the points of two dimensions, it seemed a logical transition to ask what numbers would describe the points in higher dimensions. Here we find such numbers known as quaternions, which are described as four-dimensional numbers. The question remains — do numbers stop here? As new mathematical ideas develop, applications often produce new numbers.

Look at the list in the collage. Do they all classify as complex?

---

[1]Arab texts introduced negative numbers in Europe, but most mathematicians of both the 16th and 17th centuries were not willing to accept these numbers. Nicholas Chuquet (15th century) and Michael Stidel (16th centruy) referred to negative numbers as absurd. Although Jerome Cardan gave negative numbers as solutions to equations, he considered them as impossible answers. Even Blaise Pascal, said *"I have known those who could not understand that to take four from zero there remains zero."*

[2]The Babylonian had a positional base system for writing the natural numbers using the symbols Y for 1 and < for ten, while the Egyptians did not have a positional system and would just use repetitions of the symbols | for 1 and ∩ for 10 and 9 for 100 ♀ for 1000 etc.

[3]They referred to irrational numbers as incommensurable ratios. A much related story about their origin, identifies Hippasus of the 5th century B. C. as the creator. He was at sea, and allegedly thrown overboard by Pythagoreans. The Pythagoreans were upset because his discovery refuted their contention that all things can be reduced to whole numbers or their ratios. Nevertheless, it was the Pythaogreans who proved by indirect reasoning that the $\sqrt{2}$ was irrational.

[4] Although a logical foundation had not been established for working with whole numbers, rational and irrationals and negative numbers, Hindus and Arabs freely used these numbers in their calculations. They applied positive and negative numbers as descriptive values for assets and debts. Their work was mainly engrossed in calculations, and was predominantly not concerned about its geometric validity. Consequently their arithmetic was independent of geometry.

triple junction —a mathematical occurrence in nature

In an effort to explain natural occurrences scientists and mathematicians try to discover formulae, patterns, numbers which will help them predict a natural outcome. As we know nature does not always conform, yet certain mathematical concepts have a high frequency of occurrence.

Consider the *triple junction*— an equilibrium point toward which certain natural occurrences tend.

Triple junction is essentially the point where three line segments meet. The angles at the intersection conform to 120º each.

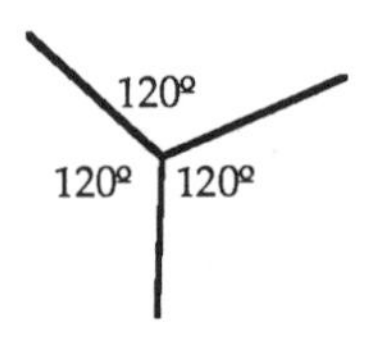

Many phenomena of nature result from the restriction caused by boundaries or availability of space, and consequently many of these result in triple junctions. First consider surface tension. Surface tension diminishes the surface area as much as possible. Consequently each soap bubble encloses a certain amount of air in such a way that the surface area for a given amount of air is minimized. This explains why a single soap bubble becomes a sphere, while a cluster of bubbles' edges, as in foam or soapy water, meet at triple junctions (three angles of 120º).

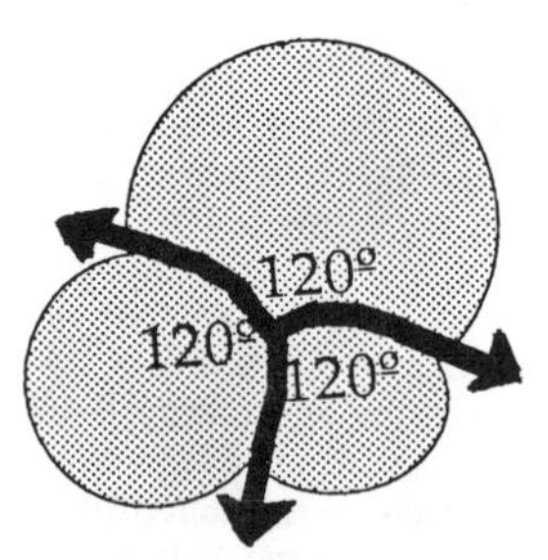

Here three soap bubbles form a cluster meeting at a triple junction.

Secondly, look at the formation of plates on a turtle's shell or look at the the manner in which bread buns(placed next to one another) expand in a pan. In each case there is a restriction created by boundary, and a close packing of these objects must take place. The shell's plates as they grow push against one another as do the bread buns in the pan when the yeast expands them. Consequently any gaps are filled and the surface contacts are minimized.

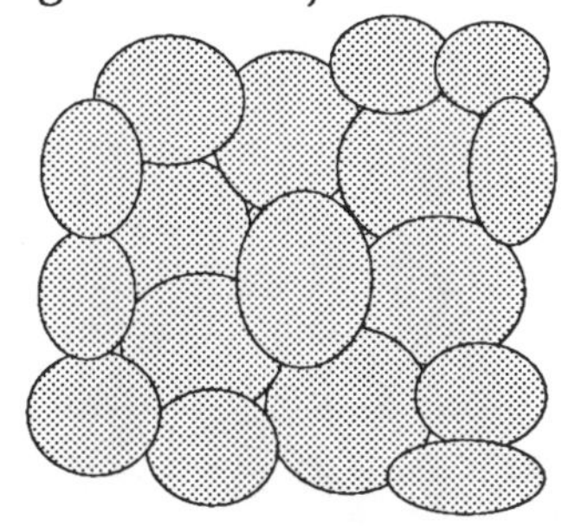

An assimilation of bread buns in a pan.

Patterns on the markings of zebra, giraffes, scales of snakes, feathers of a sparrow, scales of fish —in all we can find triple junctions occurring frequently. Other examples of triple junctions are— banana, corn , honeycomb, foam.

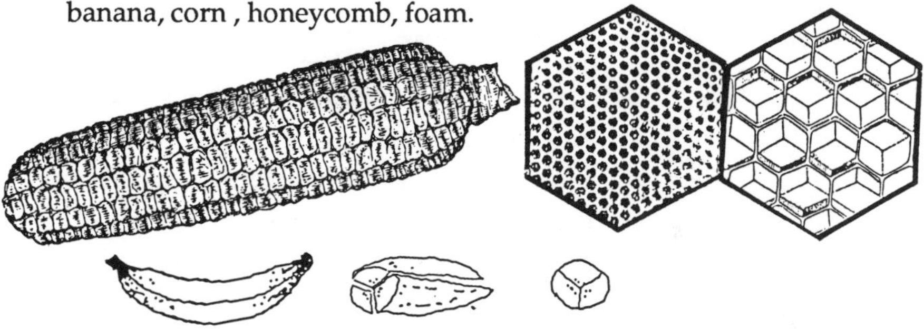

The formation of kernels on corn, the hexagonal cells of honeycomb, the interior formation of a banana are examples of triple junctions.

Another natural occurrence which results in triple junction is the cracking of elastic material such as earth or stone (as opposed to non-elastic material such as glaze). There is a tug of war in the molecular structure of drying mud. When the stress is too great the forces crack the earth into triple junctions. Here nature's restriction is to minimize the size of the opening of the crack. Cracks lay open the smallest amount of material. With wrinkles, nature's restriction is to create the minimum residue from shrinking (as in the skin of a

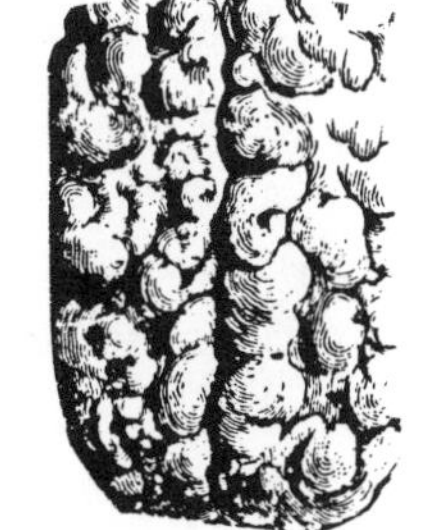

Triple junctions appear in the valleys and ridges of the human brain.

raisin) or expanding (as in the growth of the human brain). The triple junction appears in the resulting ridges or valleys of these wrinkles.

The plates of a turtle's shell form triple junctions.

Many natural occurrences can be explained by a combination of forces and circumstances which result in various forms other than the triple junction. In all these examples, nature must function under certain restrictions, be they availability of space or surface. There are various mathematical operations to find the maximum and minimum values of a function, such as using the first or second derivatives of a function; or using linear algebra to solve the maximum or minimum value of a productivity problem. Natural phenomena follow a path to minimize the work or energy in the evolution of their creations. It so happens in mathematical retrospect that the triple junction describes some of the particular works of nature.

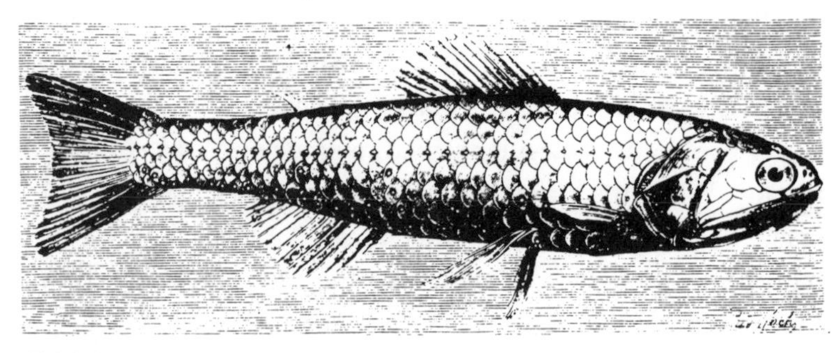

Triple junctions occur in the formation of scales on a fish.

polygonal numbers

Polygonal numbers are numbers whose shapes are connected to the shapes of regular polygons, as illustrated below.

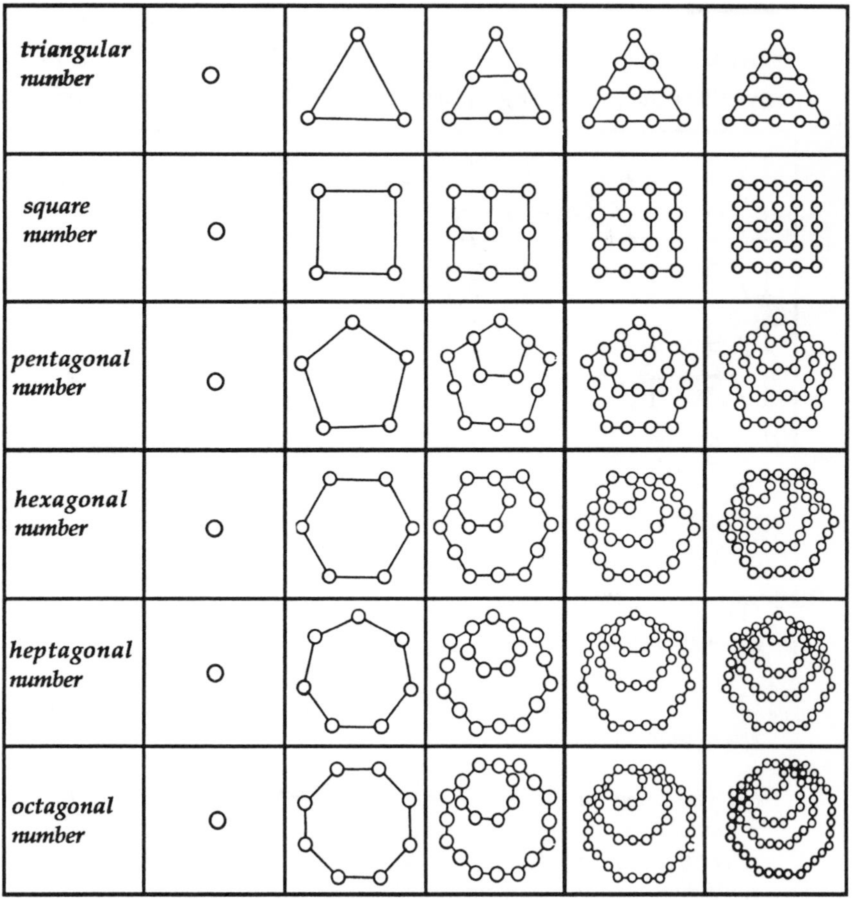

triangular number					
square number					
pentagonal number					
hexagonal number					
heptagonal number					
octagonal number					

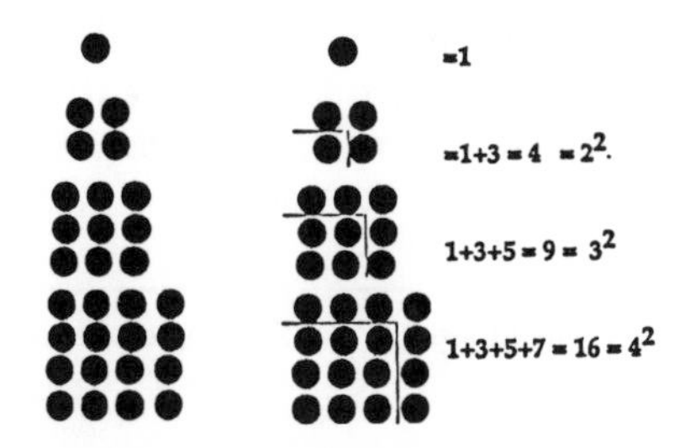

Mathematical properties have been discovered by studying the shapes that numbers form. For example, by studying the shapes of square numbers the sum of the series of odd numbers can be determined, 1+3+5+7+9+11+13…=?

the harmonic triangle

Unlike the Arithmetic (Pascal) triangle* where each term is the sum of the two terms flanking it in the row above, each term of the Harmonic triangle is the sum of all the terms in the line below it and to its right. This fact reveals some interesting information about certain infinite series.

1/1 1/2 1/3 1/4 1/5 1/6...
1/2 1/6 1/12 1/20 1/30...
1/3 1/12 1/30 1/60...
1/4 1/20 1/60...
1/5 1/30...
1/6...
.
.
.

(1) While the terms in the first line form the harmonic series which diverge, all the other series converge.

(2) The terms in the second line are one-half the reciprocals of the triangular numbers, and total 1.

(3) The terms in the third line are one third the reciprocals of the pyramidal numbers (for triangular based pyramids), and by referring to the harmonic triangle they are the terms below 1/2 and to its right and therefore must total 1/2.

(4) The terms of the harmonic triangle also possess the property that each term is the difference of the two terms above it and to its right. (e.g. 1/2=1/1-1/2) In addition, each term is also the sum of all the terms in the line below it and to its right.

---

* For additional information on the Arithmetic) (Pascal) triangle see pages 242-243

Sam Loyd's puzzling scales

This puzzle appeared in *Cyclopedia of Puzzles* (1914) by the famous American puzzlist Sam Loyd (1841-1911).

How many glasses will balance the bottle?

For the solution, see the appendix.

statistics— *mathematical manipulations*

Means, averages, medians, percent, mode, percentile, graphs... are all ways of manipulating numerical values. Take two numbers 6 and 8. One can make various comparisons— the ratio 6:8, the fraction 3/4; the percent 75%. The moment one gathers numerical values in an attempt to describe a situation, one is beginning to delve into the realm of statistics. Whether helpful or misleading, statistics is almost always influential.

Used to predict various phenomena such as—

> *a presidential candidate (Gallup poll); statistical performance (SAT scores); the status of the economy (inflation numbers, GNP growth numbers, unemployment figures, increases and declines of interest rates); the DOW (averages of the stock market); insurance rates; demographic data; weather prediction; pharmaceutical analysis on the effectiveness and/or side effects of a medication; gambling odds; incidence of ocean waves and tides—*

the domain of statistics is continually growing. When looking at the end result of any statistical analysis, one must be very cautious not to over interpret the data. Care must be taken to know the size of the sample, and to be certain the method for gathering information is consistent with other samples gathered. For example, if an exit poll in a particular election is done, the sample must be random and as large as possible. Imagine if the poll were only done in one neighborhood which was heavily weighted toward one side — a prediction based on such a small poll would be ludicrous.

Suppose a daily newspaper printed *"From a poll conducted by the Daily Enquirer, 75% of those polled had contracted flu this year."*

With this statement some people would jump to the conclusion that nearly 75% of the populace had contracted flu. The *Daily Enquirer* did not indicate the size of its poll. They may have asked only four people in their office and three of them had suffered from the flu. *No one should ever base conclusions without knowing the size of the sample and how random a sample it was.* But all too often such data is not mentioned when the statistics are given — perhaps it is overlooked or even intentionally omitted.

Another method for altering statistics is to change the make-up of your sample. For example, suppose the method for measuring unemployment was based on the number of unemployed persons in civil and private industry jobs. Now suppose after the year 1980, the statistics also include the armed services. This naturally enlarges the number of employed persons, since anyone who is in the armed services is employed. Thus, a comparison of unemployment statistics before 1980 with those as of 1980 is invalid.

With the introduction of computers and the ability to gather, sort and analyze large quantities of data quickly, statistical data and information should be more reliable as long as the agency conducting the analysis is unbiased and would not attempt to manipulate the results. The influence and power of statistics is enormous. It can be used to persuade or dissuade individuals. For example, if individuals feel their vote will not change the final result, they may not make that extra effort to vote—especially if statistics are showing exit polls leaning in one direction several hours before the polls close.

Statistics is a very powerful and persuasive mathematical tool. People put a lot of faith in printed numbers. It seems when a situation is described by assigning it a numerical value, the validity of the report increases in the mind of the viewer. It is the statistician's obligation to be aware that data in the eyes of the uninformed or poor data in the eyes of the naive viewer can be as deceptive as any falsehoods.

the mathematics of coffee & donuts

A donut and a coffee mug are topologically equivalent objects.

Both forms have exactly one hole and the mug can be transformed into the donut or vice versa by stretching, twisting and molding.

Topology studies the characteristics of objects that remain unchanged under such transformations. For example,

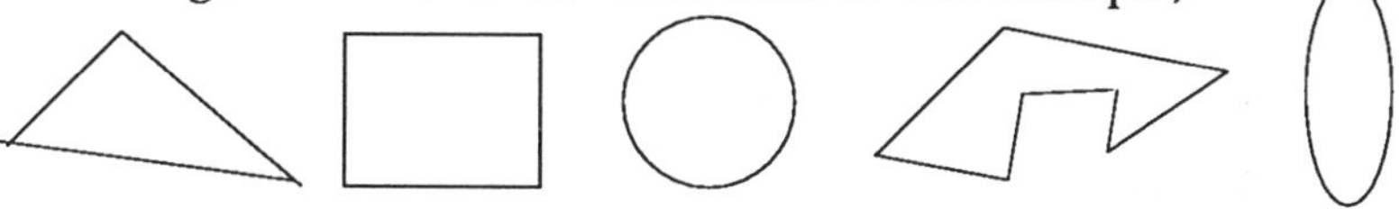

are all equivalent. In topology you consider objects as existing on a rubber sheet or plane that can be stretched or pulled. All the figures above can be transformed in this elastic geometry to the same shape, and so are equivalent. Topology does not deal with size, shape or rigid figures. In topology the characteristics of *how long, how large* are meaningless, while the properties of *where, between what, inside* or *outside* are studied. Let's see how the donut is transformed into a coffee mug.

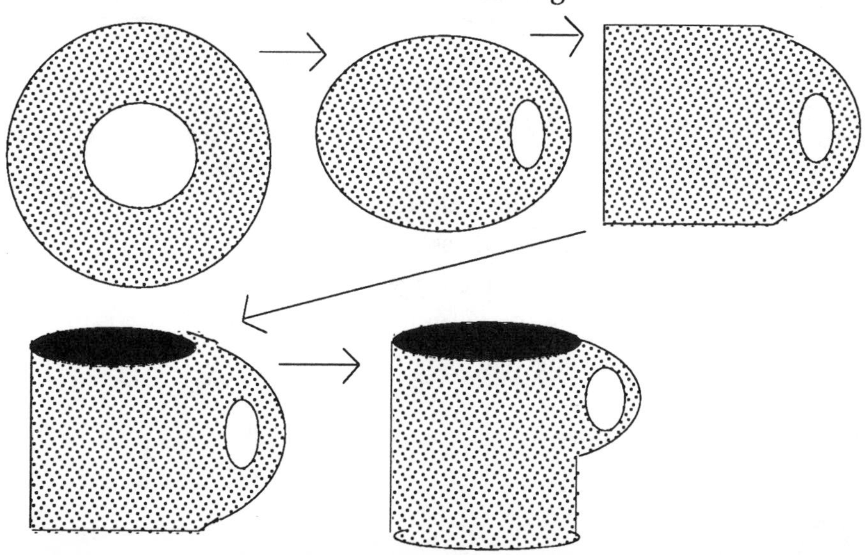

mathematical furniture

Creative minds are indeed at work when they discover everyday uses for abstract mathematical ideas. For example, the Möbius strip was used as a model to make a one-sided fan belt that wears evenly throughout—

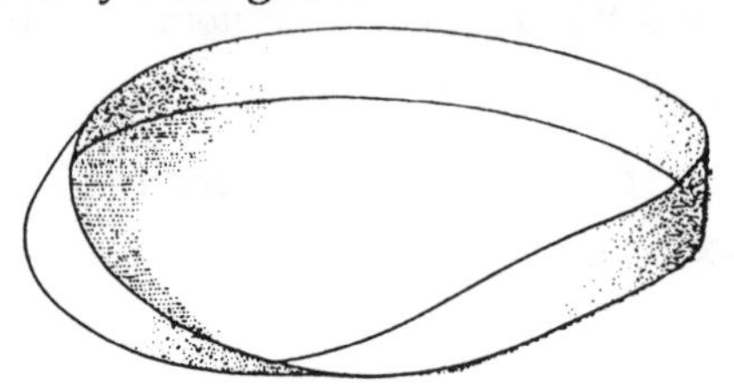

The tetrahedron was used to design a container to hold liquid drinks; fractals are generated with computers to create realistic landscape scenes.

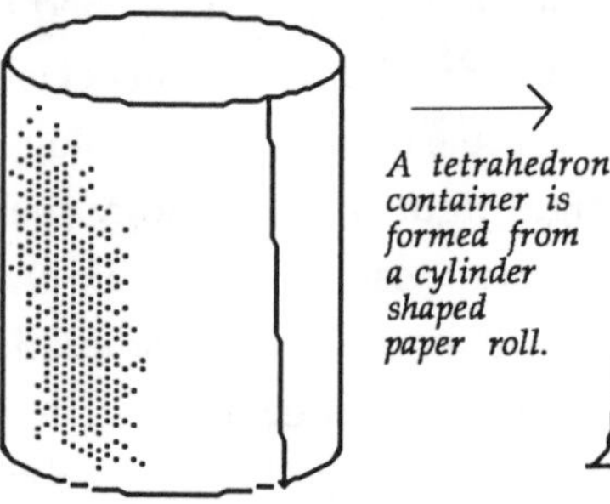

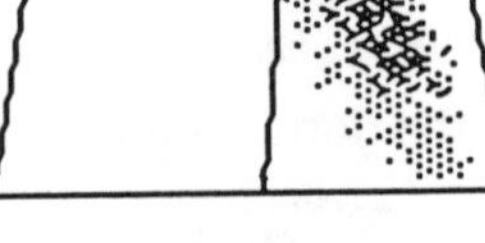

tetrahedron container

Consider the first three legged-stool. The mathematical concept that *three non-collinear points determine a plane* explains why the four-legged stool wobbles when one leg is shorter than the rest, while the three-legged stool is always stable (its three legs always rest on only one one plane).

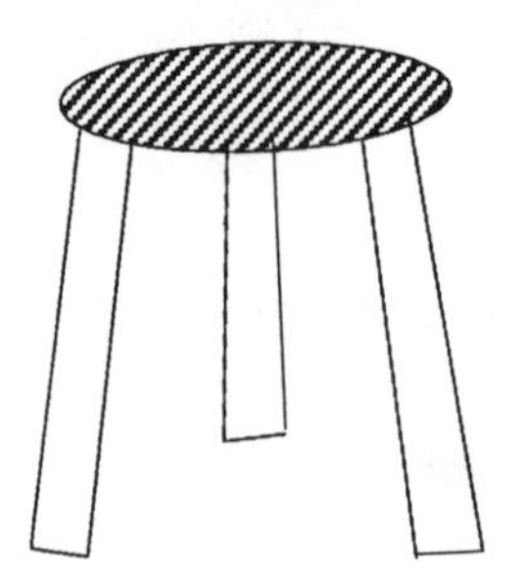

The list goes on and on.

The seven pieces of the tangram and the formation of the tangram square.

The tangram is a Chinese puzzle that became one of the most popular puzzles of the 19th century. The tangram is composed of seven pieces, of which five are isosceles right triangles, one is a square and another is a parallelogram. Using all seven pieces over 1600 designs have been created over the years. They range from the traditional tangram square to a camel, cats, birds, boats, people and a multitude of other objects.

And now Italian designer Massimo Morozzi has created the fascinating and multi-functional *Tangram Table*. His table was on display at the annual design fair — 23rd Salone del Mobile — in Milan. It is based on the seven pieces of the tangram with interestingly and varied shaped legs attached to each piece so each of the seven pieces can stand independently. Thus, all the shapes into which the tangram can be transformed, the table can also assume. It can accommodate a variety of arrangements, from a rectangular formal table setting to a whimsical cat shape.

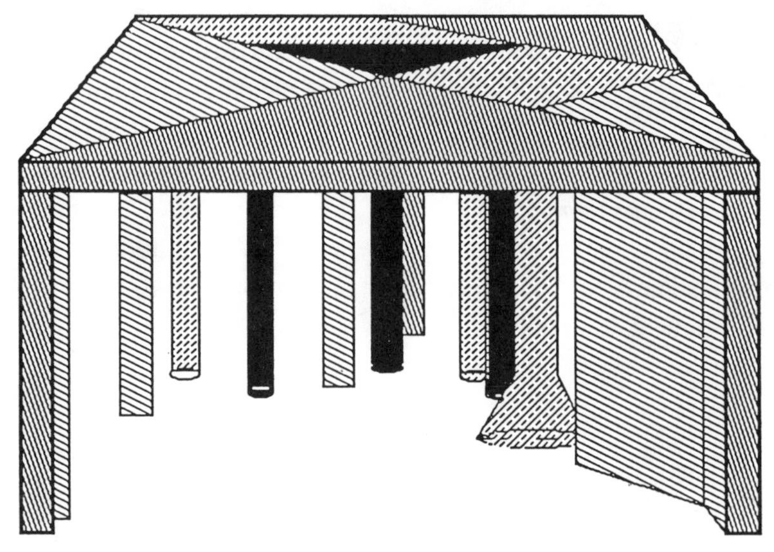

The tangram table.

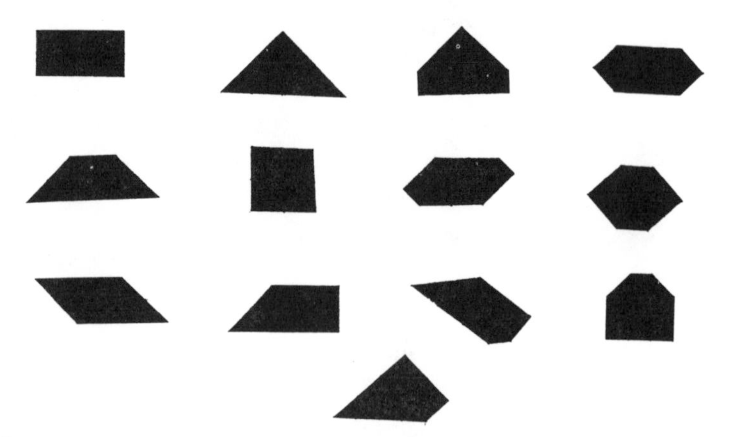

Thirteen other polygon shapes that the seven tangram pieces can form. Thus a square table can be transformed into any of these shapes. Can you determine their formation?

making rectangles

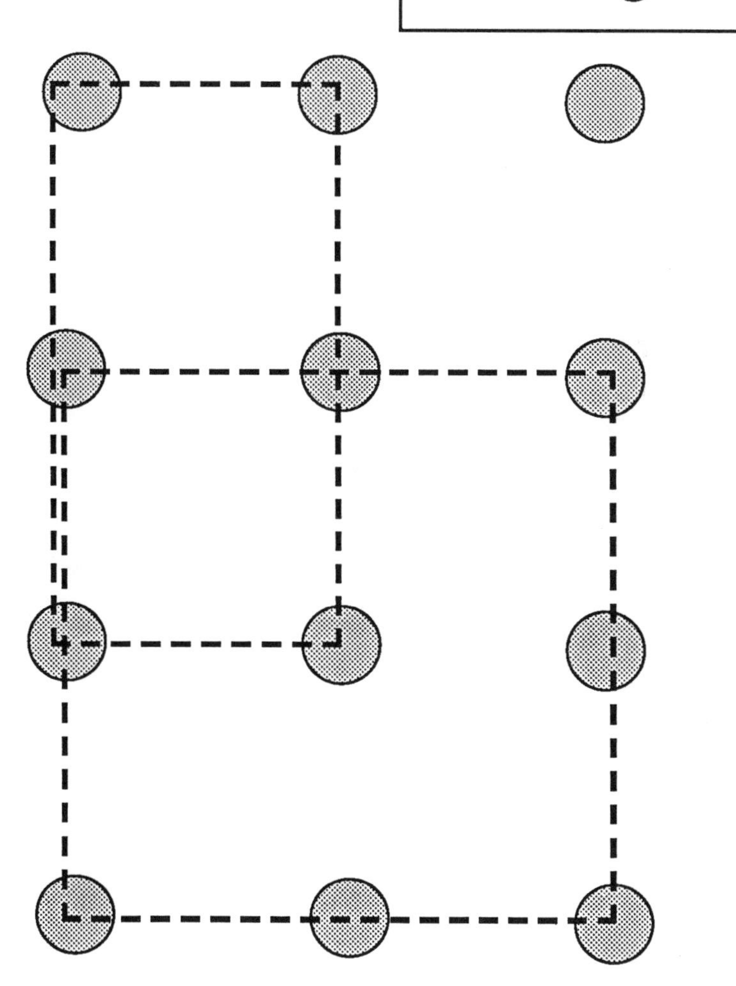

How many rectangles can be formed from these 12 dots, so that the vertices of each rectangle are 4 of the dots?

For solution, see the appendix.

geometric interpretation of prime numbers

Geometric interpretations of mathematical ideas often give another perspective and visual meaning to a concept. Prime numbers by definition are numbers other than 1 whose only factors are 1 and itself. Let's determine how to satisfy this definition geometrically.

Consider 12 squares.

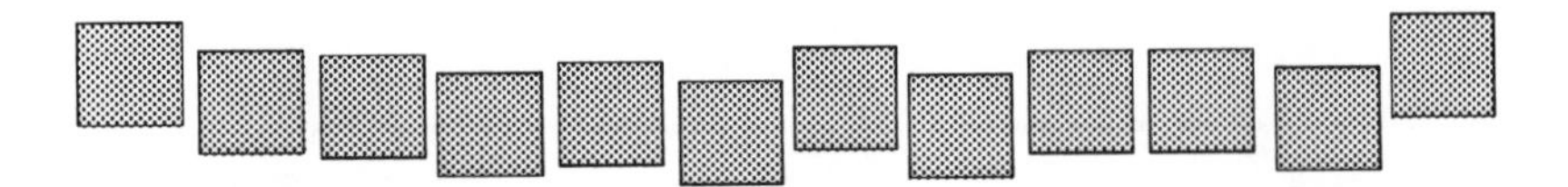

Now we'll rearrange these into various shaped rectangles.

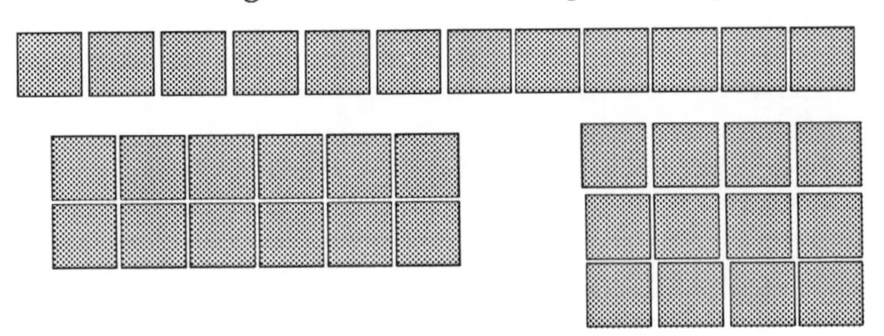

As we see, each rectangle illustrates the various factors of 12—1x12; 2x6; 3x4—the factors 1, 12, 2, 6, 3, and 4.

We now see that if a number is prime, for example, 5— it will have only one rectangle, namely

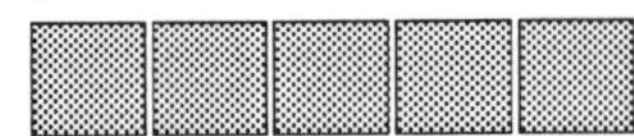

which shows 5's only factors are 1 and 5.

making an 8x8 magic square

You may think tackling an 8x8 magic square* may be a bit much. But the amount of numbers composing it does not reflect its difficulty in this case. Simply place the numbers from 1 to 64 sequentially in the square grids, as illustrated on the left. Sketch in the dashed diagonals as indicated. To obtain the magic square on the bottom, replace any number which lands on a dashed line with its compliment (two numbers of a magic square are compliments if they total the same value as the sum of the magic's square smallest and largest numbers).

1	2	3	4	5	6	7	8
9	10	11	12	13	14	15	16
17	18	19	20	21	22	23	24
25	26	27	28	29	30	31	32
33	34	35	36	37	38	39	40
41	42	43	44	45	46	47	48
49	50	51	52	53	54	55	56
57	58	59	60	61	62	63	64

64	2	3	61	60	6	7	57
9	55	54	12	13	51	50	16
17	47	46	20	21	43	42	24
40	26	27	37	36	30	31	33
32	34	35	29	28	38	39	25
41	23	22	44	45	19	18	48
49	15	14	52	53	11	10	56
8	58	59	5	4	62	63	1

*A magic square is an array of numbers arrange in a square shape in which any row, column , or diagonal total the same amount. See pages 82-86 of *The Joy of Mathematics* for more details.

the converse without the Pythagorean theorem

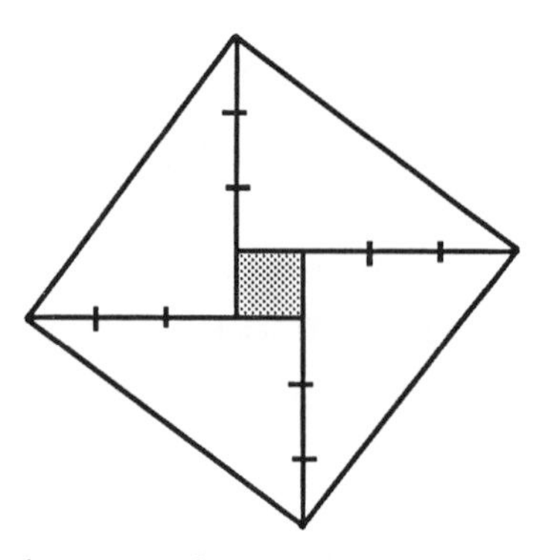

proof:
For a right triangle with legs 3 and 4, the hypotenuse is 5.
Take 4 right triangles with legs 3 and 4 , and form the square shown above. The sum of the areas of the four triangles is 24 square units with the interior square's area being 1 square unit. This gives the area of the large square as 25 square units, which shows the sides of the large square to be 5 each. Therefore, the hypotenuse of a right triangle with legs 3 and 4 must be 5, and $3^2 + 4^2 = 5^2$.

This elegant proof appeared in the Chinese book *Chou-pei Suan King* (circa 300B.C.). It is especially interesting since it proves an example of the *converse* of the Pythagorean theorem without using the Pythagorean theorem. Beginning with a right triangle having legs 3 and 4, a square is formed using four congruent triangles. Although the proof specifically illustrates the 3-4-5 right triangle, its method can be generalized for any right triangle with legs of length a and b; and is a great way to show the converse of the Pythagorean theorem.

To generalize the proof, label the legs in the diagrams as a and b and the hypotenuse as c. Calculate the area of the square in two ways:

(1) the sum of the four triangles + the interior square =
(2) area of the large square

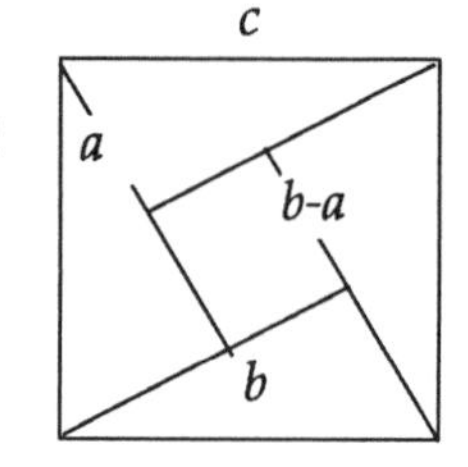

$$(1/2)a \bullet b + (1/2)a \bullet b + (1/2)a \bullet b + (1/2)a \bullet b + (b-a)^2 = c^2$$
$$2(a \bullet b) + (b-a)^2 = c^2$$
$$a^2 + b^2 = c^2$$

For steps 1-3 refer to diagram 1

1) Take any ΔABC.
2) Bisect ∠B with ray BD.
3) Construct the perpendicular bisector of segment AC. Label the point, E, where ray BD intersects the perpendicular bisector of AC. Label the point, F, where the perpendicular bisector intersects segment AC.

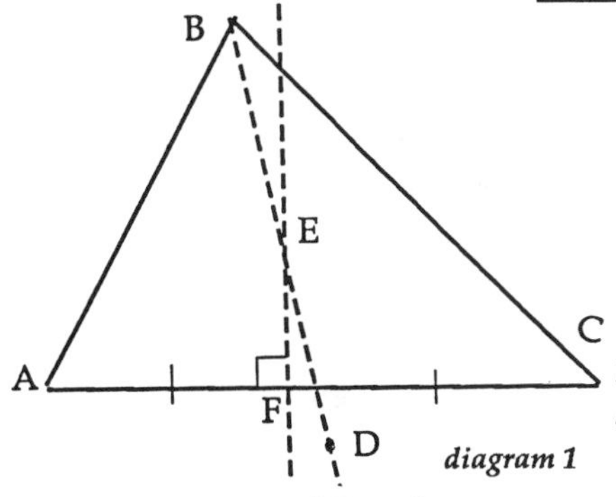

diagram 1

For steps 4-12 refer to diagram 2.

4) From point E construct a perpendicular segment to segment AB and one to segment BC. Label these segments EG and EH respectively.
5) Add segments AE and EC to the diagram.
6) ΔBGE≅ΔBHE by SAA— ∠ 1≅∠ 2 because ∠B is bisected, step 2; ∠ G=∠ H because both are right angles, step 4; |BE|=|BE|, identity
7) |GE|=|HE| because ΔBGE≅ΔBHE, step 6
8) |AE|=|EC| because line EF is the perpendicular bisector of segment AC, thus any point on line EF is equidistant from points A and C.
9) ΔAGE≅ΔCHE because they are right triangles with a pair of congruent legs and congruent hypoteni, steps 7 and 8.
10) (a)|AG|=|CH| because they are congruent parts of congruent triangles, step 9.
(b) |BG|=|BH| because they are congruent parts of congruent triangles, step 6.
11) |AB|=|BC| step 10, addition of equals

∴ 12) ΔABC is isosceles, step 12.

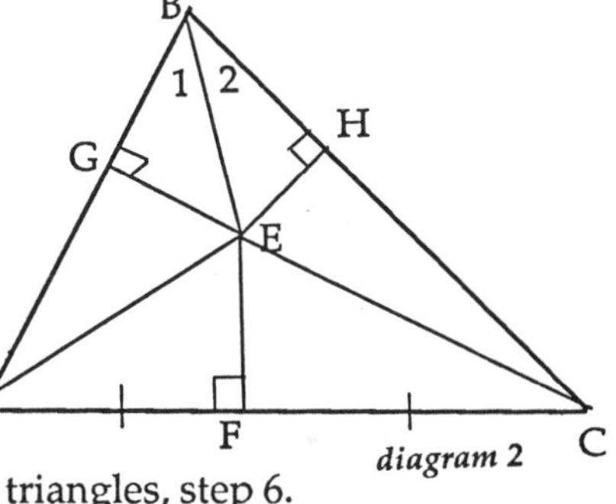

diagram 2

See appendix for explanation.

in search of perfect numbers

6, 8, 496, ?

Pythagoreans believed that whole numbers were the essence of everything. They even personified certain numbers. For example, even numbers were female. The Pythagoreans studied properties and types of numbers. *Perfect numbers* were one of the types of numbers on which they focused. These are numbers which are equal to the sum of their proper divisors. 6 is an example of a perfect number because its proper divisors, 1, 2, and 3 total 6. 28 and 496 are also examples of perfect numbers. The last theorem in volume nine of Euclid's *Elements* deals with perfect numbers. It states:

If 2^n-1 is a prime number, then $2^{n-1}(2^n-1)$ is a perfect number.

For n=2 we get the perfect number 6. For n=4, 2^4-1 is not prime, so the conclusion would not yield a perfect number. The search for perfect numbers has tantalized and perplexed mathematicians for centuries.

Until now no one has found an *odd perfect number,* and no one has shown that one does not exist*, which is the converse of Euclid's theorem—

Every perfect number is of the form $2^{n-1}(2^n-1)$, with 2^n-1 a prime.

As of yet, no one has proven this theorem. The Swiss mathematician, Leonard Euler (1707-1783), proved all even perfect numbers are of this form. The search for perfect numbers continues today. With the aid of computers, perfect numbers were found for n=521, 607 1279, 2203, 2281, 3217, 7090, 4253, 4423; and these are the only ones for n<5000. Perfect numbers are also given by n=9689, 9941, 11213, 19937. You can imagine how large these perfect numbers are. For example, in 1963, the University of Illinois mathematics department discovered the perfect number for n=11213. It contains 6751 digits and has 22425 divisors.

---

*This is one of the famous unsolved problems of number theory.

$\sqrt{2}$ *dynamic rectangle*

Dynamic rectangles[1] have a tendency to appear in many aspects of art. Here we see how a $\sqrt{2}$ rectangle surrounds the shape of this ancient food bowl from Pueblo Viejo in upper Gila Valley[2].

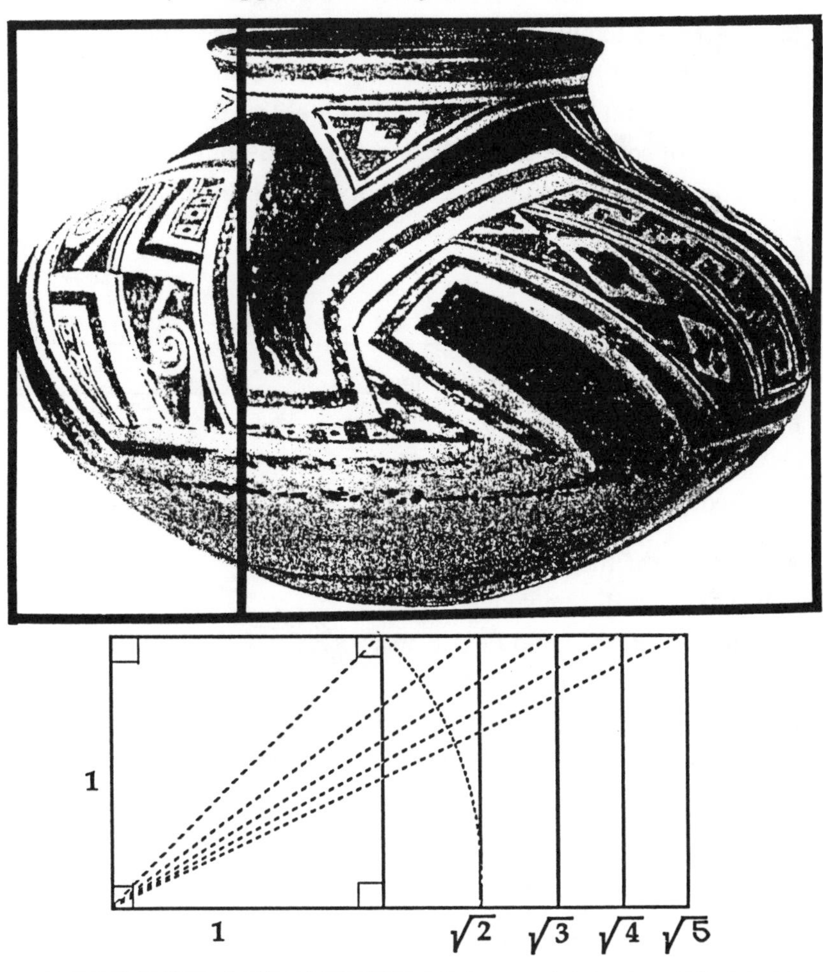

---

[1] Dynamic rectangles are rectangles generated from the unit square.

[2] Reprinted with permission of Dover Publications, *Authentic Indian Designs* edited by Maria Naylor.

flipping over the Möbius strip

Although it has been 143 years since the Möbius strip was created by German astronomer/ mathematician Augustus Ferdinand Möbius (1790-1868), its properties still amaze and stimulate the imagination. In 1858 Möbius presented his strip as a simple single-sided single-edged phenomenon. Over the years mathematicians, artists, scientists, writers have put the Möbius strip through its paces. Here is a chronicle of some of their work and findings. But the best way to enjoy this wonderful object's properties is by making your own models.

The Möbius strip & properties

•The Möbius strip is formed when a rectangular strip of paper is given a half-twist and its ends attached.

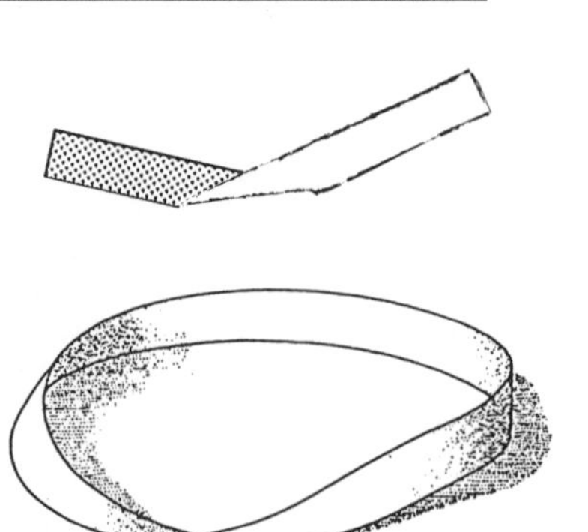

•Test to see if the Möbius strip is one-sided and one-edged by tracing along the middle of its surface with a pencil. Do you return to your starting point without lifting your pencil? Make a starting notch on its edge and run your finger along its edge. If you trace the entire edge and return to the notch without lifting your finger, it has only one edge.

•Using a scissors, cut the Möbius strip along the center line. Do you end-up with one loop? If so, is the new loop 2-sided and 2-edged? Test it by cutting it again. If you get one loop, it's not 2-sided[1].

•Take a long rectangular strip and color a thick line down its middle on both sides of the strip, so that the color covers a third of its width. Now twist this strip into a Möbius strip. Using a scissors, cut the Möbius strip along the edge of the colored line (i. e. one-third of the way from the edge of the Möbius strip). What did you get?[2]

•Envision a Möbius strip world with the strip made from a clear surface. Creatures live in a thin nearly transparent film of the strip. Suppose a cat of this world, which would appear as a 2-D silhouette, begins to walk from a point inside the strip until it returns to this point. Reaching its starting point, the cat will find itself flipped over, as if it had changed directions. What happens if it goes around again? If a 2-D right-handed glove would go along the same path as the cat, it would have become left-handed when it reached the starting point. This phenomenon is analogous to a 3-D right-handed glove becoming a left-handed glove via the 4th dimension.

• Place two rows of two dots on 2-D plane. All pairs cannot be connected without some segments crossing. Place these dots on a Möbius strip and see what you can do.

• Make a strip with two half-twists, and test if it is a Möbius strip. Cut it along its center. Do you end up with one or two loops? How many half-twists does each have? [3]

• Make a strip with three half-twists, and test if it is a Möbius strip. Cut it along its center, and end up with one loop with a knot in it. This knot will be a trefoil knot.

•Make a "double" Möbius strip by simultaneously taking two identical rectangular strips and giving them a half-twist. Then join respective ends together. Is this actually two Möbius strips nestled together? Run your finger between them.

—Unnestle the "double" loop. You should find that you have one loop with four half-twists.

—Cut both parts down the middle at the same time. You will end up with two linked loops.

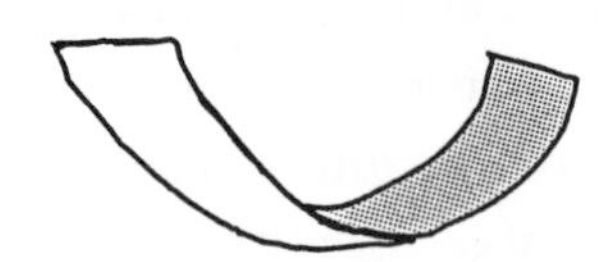

This strip of paper as this is two-sided and single-edged.

—Make a new "double" Möbius strip and begin cutting along the center of the top loop; or trace a line down its center with a pencil. You will have two loops linked.

This band is two-sided and two edged.

—The edges of a "double" Möbius strip appear to be parallel and distinct. Are they?

—Some feel the "double" Möbius strip can be used to gain access to a higher dimension because a walk once around between them will transport you to the same starting point but you will be upside down. Suppose a spider begins to crawl along the "floor" of the "double" Möbius strip. When it reaches its starting point it will be on the "ceiling", and the only way it can get back to the floor is to go around again.

•If two Möbius strips are joined along their edges lengthwise, they will form the famous Klein bottle. Or if a Klein bottle is split lengthwise, two Möbius strips are formed.

•The Möbius strip comes in right and left-handed strips that are mirror images of one another.

Practical Applications

•Möbius molecule — In 1981 David M. Walba of the University of Colorado synthesized a molecule in the form of a Möbius strip. It was a double-ladder strip of carbon and oxygen atoms. Scientists are now exploring the mathematics of knots and links while working with molecules.

•B. F. Goodrich Co. has a patent for a Möbius strip conveyor belt. It lasts longer since the wear and tear is spread uniformly over the entire surface.

•A Möbius film strip that records sound on both sides was devised by Lee De Forest in 1923. The same idea has been used with tape recorders.

• O. H. Harris patented a Möbius abrasive belt.

• J.W. Jacobs made a self cleaning Möbius filter belt for a dry cleaning machine in 1963.

•Richard L. Davis invented a nonreactive resistor Möbius strip, and is patented by the U.S. Atomic Energy Commission.

The Mobius strip in the Arts

•A Möbius steel strip sculpture is at the Museum of History and Technology at the Smithsonian in Washington D.C.

•M.C. Escher used the Möbius strip in his woodcuts, *Möbius strip I* and *Möbius strip II*

•The Möbius strip has been a central theme in sculpture, a *New Yorker* magazine cover, postage stamps, and graphic art logos.

•The Möbius strip has been used in many stories. *Time Squared,* an episode of *Star Trek, The Next Generation* , used the idea of a Möbius strip.

---

[1]If the original loop has n-odd number of half-twists, its bisecting results in a loop with 2n+2 half-twists. If the original loop has an even number of half-twists, its bisecting produces 2 separate narrower loops, yet identical to the first loop.

[2]It results in two loops one colored and half the length of the original strip and other the same length as the original Möbius strip. If these bands are nestled so it appears as three loops, the colored loop seems to separate the other "two". This can also be made by taking 3 strips the same size, give them a half-twist and join their respective ends.

[3]Two loops each with two half-twists

Ovid's game

In addition to spanning a wide variety of different subjects, topics, and areas of influence, mathematics is a way of thinking. Here is a game to play and analyze mathematically.

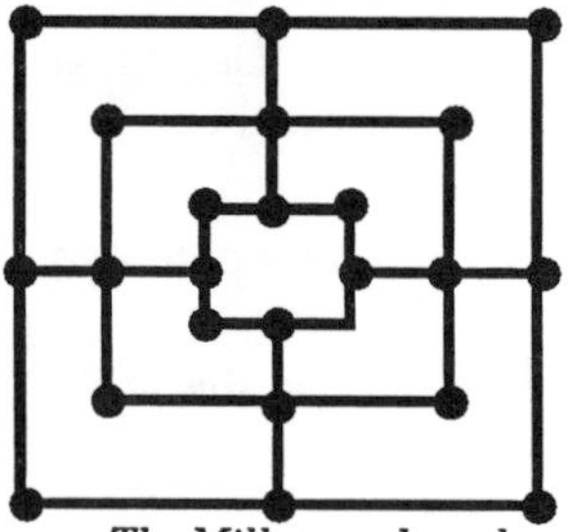

The Mill game board.

We are all familiar with the game of Tit Tat Toe. This is a variation of that game, which dates back to the works of the versatile Roman poet, Ovid (43 B.C.-17 A.D.). Ovid's game and variations of its board (e. g. the Mill) have been found on the steps of the Acropolis in Athens, on Etruscan pottery, and Roman tiles. In English literature it is known as *Nine Men's Morris* (possibly because of the resemblance to a Moorish dance). William Shakespeare (1564-1616) makes reference to *Nine Men's Morris,* in *A Midsummer Night's Dream,* where it is a game played on a square in the turf marked with nine stones.

"The nine men's morris is filled up with mud"
Act II, Scene 2. line 98

This illustration of people playing nine men's morris is from the game book of Alfonso X of Spain.

Ovid's Game

This is a game for two players.

- Each player has three playing pieces.
- The object of the game is to align your three pieces in a row.
- Players alternate turns placing their pieces.
- If no player has aligned his or her three pieces each player moves to an unoccupied adjacent spot until someone gets his or her three pieces in a row.

After you have developed a strategy for these rules, add the following restriction and see what transpires:

- No player's opening play may be the center.

The Ovid game board

This particular game shows the two players having placed their pieces without either aligning their three pieces. At this point players begin to move their pieces along the lines until one player is successful in aligning his or her three pieces.

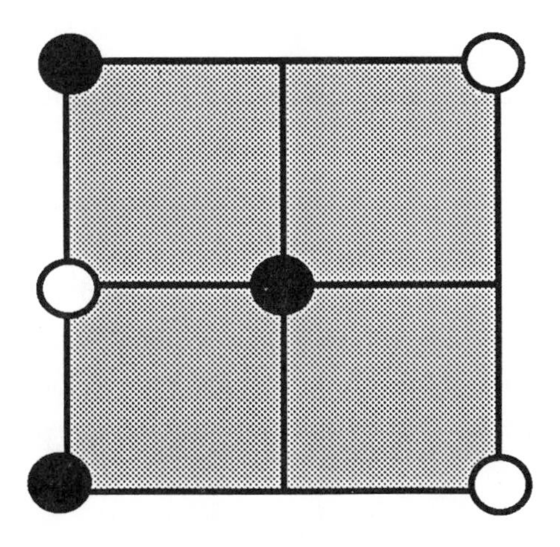

number patterns of the Stone Age

These markings could represent doodles or designs by an artist drawing on a cave some 25,000 years ago. Yet on the other hand they may be some of the first representations for numbers.

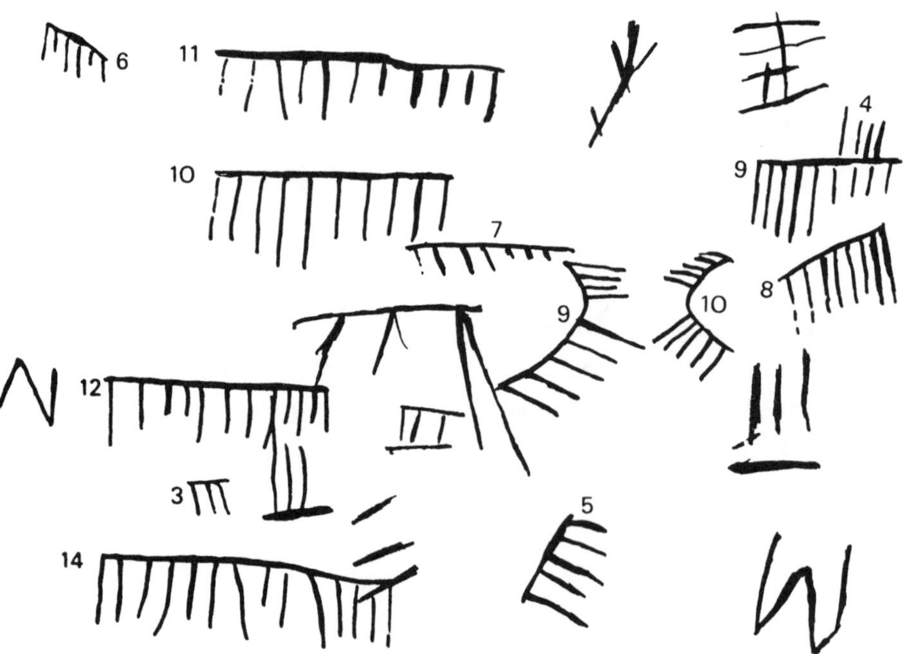

Stone Age number patterns found in La Pileta, Spain.

La Pileta Cave in southern Spain was inhabited over 25,000 years ago. Other drawings appear in the caves and artifacts have been discovered. The skeletons and weapons that were found indicate that the caves were still inhabited in the Bronze Age —1500 B.C. The ceramic pieces that were discovered are believed to be some of the oldest known pottery remains in Europe, dating back to the Neolithic period (the 3rd millennium).

the nine point circle

The nine point circle is a geometrical concept that links some of a triangle's parts that one would initially feel could not be interconnected. This circle was discovered by Charles Julien Brianchon (1783-1864) and Jean Victor Poncelet (1788-1867).

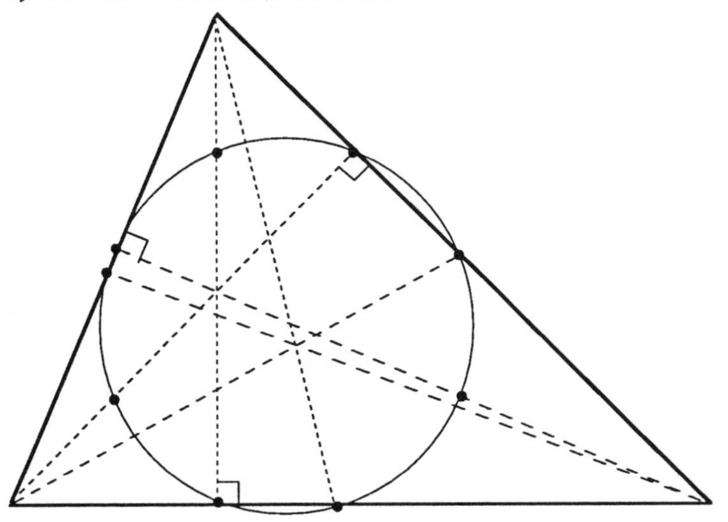

They found that —
every triangle has a circle which passes through the 3 endpoints at the bases of the altitudes of the triangle, through the 3 midpoints of the sides of the triangle, and through the 3 midpoints of the segments whose endpoints are the orthocenter (point where the altitudes of a triangle intersect) and the vertices of the triangle.

Then in 1822 K. W. Feuerbach proved a fascinating relationship between a triangle's nine point circle and its incircle and excircles. He proved that the nine point circle is tangent to the triangle's incircle and also to each of the triangle's three excircles*.

*An incircle is the circle inscribed inside a triangle, and an excircle is the circle of a triangle tangent to one of its sides and also to the two extended sides. See diagram.

architecture & mathematics

For thousands of years mathematics has been an invaluable tool for design and construction. It has been a resource for architectural design and also the means by which an architect could eliminate the trial and error techniques of building. Here is a partial list of the some of the mathematical concepts used over the centuries:

- pyramids
- prisms
- golden rectangles
- optical illusions
- cubes
- polyhedra
- geodesic domes
- triangles
- Pythagorean theorem
- squares, rectangles, parallelograms
- circles, semicircles
- spheres, hemispheres
- polygons
- angles
- symmetry
- parabolic curves
- catenary curves
- hyperbolic paraboloids
- proportion
- arcs
- center of gravity
- spirals
- helices
- ellipses
- tessellations
- perspective

The design of a structure is influenced by its surroundings, by the availability and type of materials, and by the imagination and resource upon which the architect can draw. Some historical examples are—

- The task of computing size, shape, number and arrangement of stones for the construction of the pyramids of Egypt, Mexico and the Yucatan relied on knowledge of right triangles, squares, Pythagorean theorem, volume and estimation.
- The regularity of the design of Machu Picchu would not have been possible without geometric plans.
- The construction of the Parthenon relied on the use of the golden rectangle, optical illusions, precision measurements

The use of glass, varied angles and shapes make this building (Foster City, California) appear to change at different times of day and as viewed from various locations. It is fascinating to watch it reflect its surroundings.

and knowledge of proportion to cut column modules to exact specifications (always making the diameter 1/3 of the height of the module).

- The geometric exactness of layout and location for the ancient theater at Epidaurus was specifically calculated to enhance acoustics and maximize the audience's range of view.
- The innovative use of circles, semicircles, hemispheres and arches became the main mathematical ideas introduced and perfected by Roman architects.
- Architects of the Byzantine period elegantly incorporated the

concepts of squares, circles, cubes and hemispheres with arches as used in the St. Sophia church in Constantinople.

- Architects of Gothic cathedrals used mathematics to determine the center of gravity to form an adjustable geometric design of vaulted ceilings meeting at a point that directed the massive weight of the stone structure back to the ground rather than horizontally.
- The stone structures of the Renaissance showed a refinement of symmetry that relied on light and dark and solids and voids.

With the discovery of new building materials, new mathematical ideas were adapted and used to maximize the potential of these materials. Using the wide range of available building materials—stone, wood, brick, concrete, iron, steel, glass, synthetic materials such as plastic, reinforced concrete, prepounded concrete — architects have been able to design virtually any shape. In modern times we have witnessed the formation of the hyperbolic paraboloid (St. Mary's Cathedral in San Francisco), the geodesic structures of Buckminster Fuller, the module designs of Paolo Soleri, the parabolic airplane hanger, solid synthetic structures mimicking the tents of the nomads, catenary curve cables supporting the Olympic Sports Hall in Tokyo, and even an octagonal home with an elliptical dome ceiling. Architecture is an evolving field. Architects study, refine, enhance, reuse ideas from the past as well as create new ones. In the final analysis, an architect is free to imagine any design so long as the mathematics and materials exist to support the structure.

the I Ching & the binary system

The I Ching or Oracle of Change is one of the oldest books in the world. It represents an ancient Chinese philosophy which encompasses a psychology, a calendar and an oracle. Although its early origins are unknown, it probably dates back to the 8th century B. C. The hexagram—six horizontal segments (with the solid segments being the Yang and the broken segments the Yin)—is its archetypal structure. 64 hexagrams form the complete series of permutations.

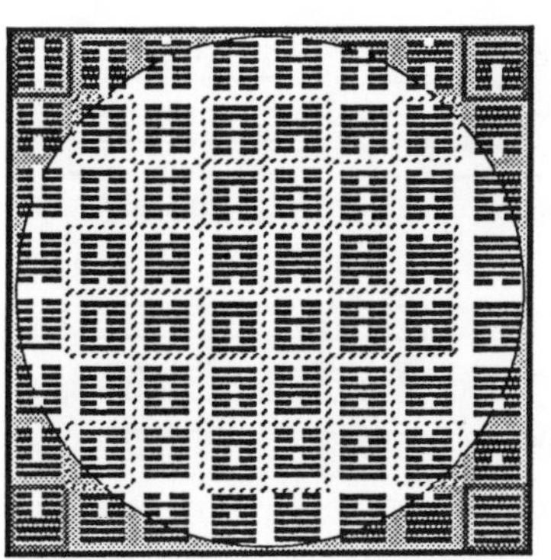

Mathematician, scientist, philosopher, linguist and diplomat Gottfried Wilhelm Leibniz (1646-1716) first wrote about the binary system in his paper *De Progressione Dyadica*, 1679. From 1697 to 1702 he corresponded with Père Joachim Bouvet, a Jesuit missionary in China. It was through Bouvet that Leibniz learned that the I Ching hexagrams were connected to his binary numeration system. He noticed that if he replaced zero for each broken line and 1 for the unbroken line, the hexagrams illustrated the binary numbers.

For example:, taking the hexagrams in order we find:

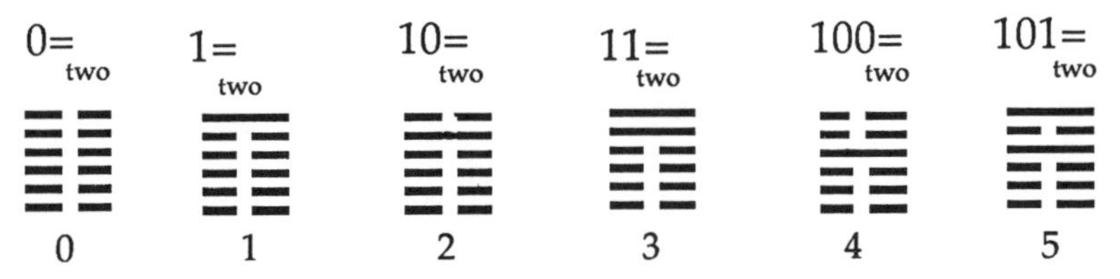

Although Leibniz and Bouvet felt that the Chinese had discovered the binary system through I Ching, there is no evidence to indicate this.

music of the spheres

In ancient Greek times, mathematics, music and astronomy shared a common curriculum of study. The Pythagoreans linked numbers and musical scales. Based on their knowledge of mathematics, music and the orbits of planets, they formulated an idea referred to as *Music of the Spheres,* which linked music and astronomy.

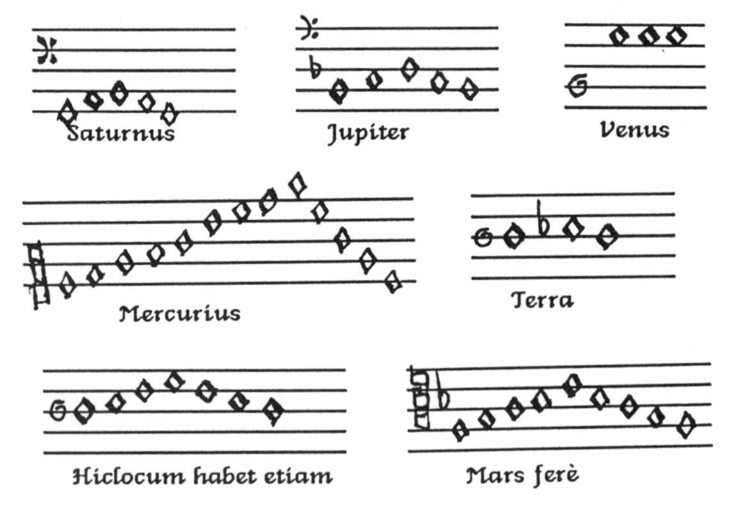

Johannes Kepler (1571-1630) felt a need to find some order and interrelationships in the concepts with which the universe conformed. In 1618 he published *Harmony of the World.* He linked the velocities of planets in their elliptical orbits with musical harmony.* He related the greatest and least velocity of a planet to the musical scale. He considered this one of his greatest achievements—the apotheosis of the *Music of the Spheres* of the ancient Greeks.

---

* Today no scientific significance is attached to this astronomy-music relationship. Yet his work in this area led to many valuable astronomical discoveries, e.g. Kepler calculated the speed of the Earth in its elliptical orbit.

anamorphic art

Anamorphic art is the art of distorting images. In a sense one could call one's shadow a type of anamorphic art, since one's shadow is a distortion of one's actual body shape. This type of art was created both for amusement and for concealing objects. For example, during the reigns of Kings George I and II an anamorphic portrait of the 'pretender' to the throne, Charles Edward Stuart, may have been used by supporters to conceal their allegiance. This art was also used to carry political protest messages, or with erotic or pornographic illustrations. In addition, anamorphic paintings were sold with anamorphoscopes (in the shapes of cylindrical or conical mirrors) as toys in the 18th and 19th centuries.

a distorted scorpion

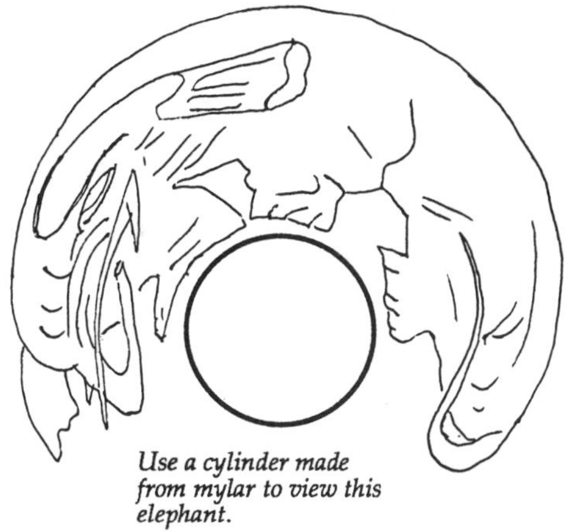

Use a cylinder made from mylar to view this elephant.

Techniques of projective geometry are used to create anamorphic drawings. The artist distorts the image through transformations either to a distorted plane, or onto a cylinder or conical tube. Frenchman Jean François Niceron demonstrated his geometric technique for creating anamorphic art in his book (1638), *La Perspective Curieuse.* It is not unlikely that anamorphic art will experience a renaissance, since distortions can be easily created using computer graphics.

measurement problem

If you were allowed four fixed lengths, what length straightedges would you choose to measure the integral distances from 1 unit to 40 units? You cannot use a straightedge more than once to measure a desired distance.

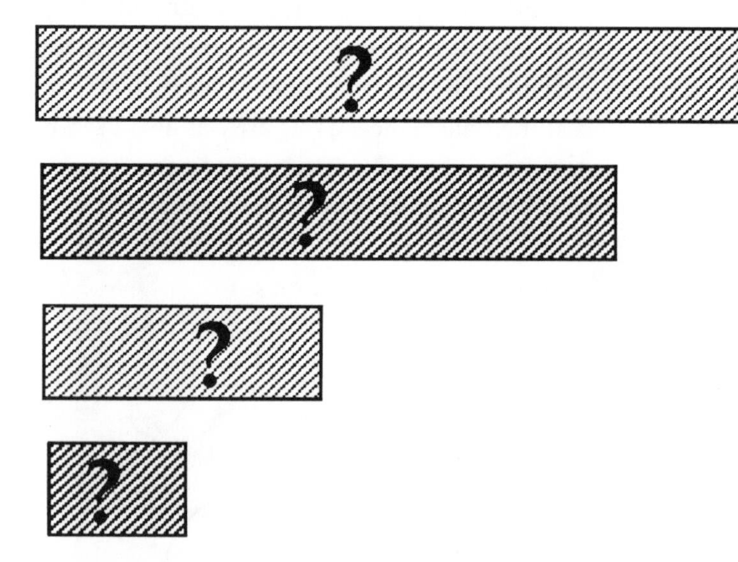

For example, if you choose 1 unit, 2 unit and 4 unit straightedges, the largest distance you could measure would be 7—

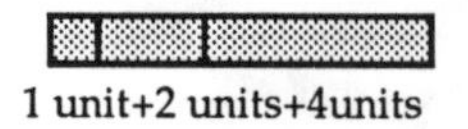

The distance 3 units could be measured in two ways—

What four sized straightedges would you choose to measure the whole number lengths from 1 to 40 units?

For solution, see the appendix.

a Renaissance illusion

Projective geometry is a field of mathematics that deals with properties and spatial relations of figures as they are projected — and therefore with problems of perspective. To create their realistic 3-D paintings, Renaissance artists used the now established concepts of projective geometry — point of projection, parallel converging lines, and the vanishing point. 18th century artist, William Hogarth's print *False Perspective* has the comment *"Whoever makes a design without the knowledge of perspective will be liable to such absurdities as are shown in this frontilpiece. "* How many perspective flaws can you find in Hogarth's work?

Whoever makes a DESIGN without the Knowledge of PERSPECTIVE will be liable to such Absurdities as are shewn in this Frontispiece.

inversions

Symmetry abounds in nature and mathematics. Be it revealed in the shape of a maple leaf or that of a hexagon, symmetry is an innate part of many forms. With inversions we discover a world of objects. The forms of letters and words do not necessarily possess symmetry at their inception, but with the ingenuity of Scott Kim, are transformed into fascinating shapes and styles. An inversion is a word that is written so that it possesses one or more types of symmetries. As such, mathematical ideas are presented in the shapes and positions of letters and words, and we discover that inversions can even enhance the meanings of the words.

©1989 by Scott Kim. Reprinted from his book *Inversions* published by W.H. Freeman.

Scott Kim is a graphic designer, computer specialist and mathematician. Isaac Asimov describes him as "the Escher-of-the-Alphabet".

Study this inversion. You should be able to eventually see the word upside-down. Turn the page upside-down. What do you see? ©1989 by Scott Kim.

Place a mirror vertically along the "i". ©1989 by Scott Kim.

Romeo & Juliet puzzle

The puzzle *Romeo & Juliet* appeared in *Canterbury Puzzles* (1907) by the renowned English puzzlist Henry Dudeney (1847-1930).

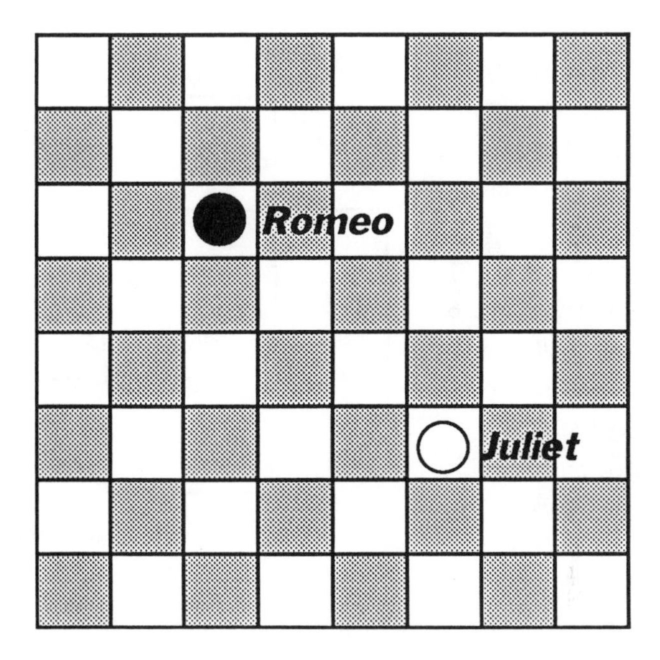

Romeo must make his way to Juliet. He must pass through each of the squares only once before reaching Juliet. And he must reach Juliet in the fewest possible turns.

For the solution, see the appendix.

What do you "mean"?

There is an ancient tie between arithmetic and harmonic means. While the Pythagoreans were studying the relationships between musical tones and ratios of numbers, they discovered that if a string of length 12 was shortened to 3/4 its length it produced the fourth of the original tone, 2/3 its length it produced the fifth, and 1/2 its length gave its octave. These were represented by the numbers 12, 9, 8, and 6. Surprisingly, these numbers were also linked by their arithmetic and harmonic means, namely — 9 is the arithmetic mean of 6 and 12; while 8 is the harmonic mean of 6 and 12. In addition, the Pythagoreans called the cube a harmonic body because 6, 8, and 12 represented its faces, edges and vertices respectively.

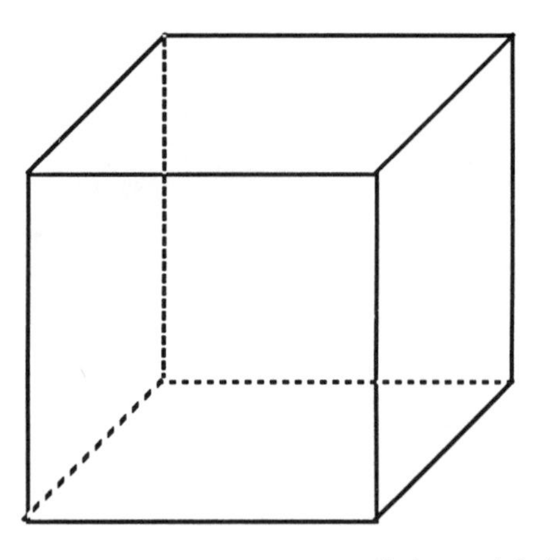

the harmonic body

* * *

Let's look at the different uses of the word *mean* in mathematics.

Pythagoras defined the *arithmetic mean* as a number that exceeds the first number by the same amount as the second number exceeds the mean. So we see in order to find the arithmetic mean, we must find the number that is midway between the two given numbers, *(a+b)/2 =arithmetic mean of a and b.* Today we often call this mean the *average.* —

To find the ***arithmetic mean*** for $a_1, a_2, a_3, \ldots, a_n$,

we find the sum of the terms
and divide by how many terms we totaled

$$(a_1+a_2+a_3+\ldots+a_n)/n.$$

The *geometric mean, c,* of a and b was defined as $(c-a)/(b-c) = a/c$. Today it's defined equivalently as $c=\sqrt{a \bullet b}$ or $a/c = c/b$, and is found in various geometric problems as in the altitude to the hypotenuse of a right triangle and the golden mean in the golden rectangle.

To find the ***geometric mean*** for $a_1, a_2, a_3, \ldots, a_n$,

we find the nth root of their product, namely

$$\sqrt{a_1 a_2 a_3 \ldots a_n}.$$

The *harmonic mean, H* was defined as $(H-a)/(b-H) = a/b$. Today it is defined as $1/H = (1/a + 1/b) / 2$. Another way to think of it is as the *reciprocal of the arithmetic mean of the reciprocals of the numbers.*

The harmonic mean for $a_1, a_2, a_3, \ldots, a_n$ is

$$1/[(1/a_1+1/a_2+1/a_3+\ldots+1/a_n)/n].$$

mathematical connections

It is not surprising to find many mathematical ideas interconnected or linked. The expansion of mathematics depends on previously developed ideas. The formation of any mathematical system begins with some undefined terms and axioms (assumptions) and proceeds from there to definitions, theorems, more axioms and so on. But history points out this is not necessarily the route that creativity takes. For example, Euclidean geometry did not begin with Euclid's book, *The Elements.* Instead Euclid wrote his book after studying, compiling, and organizing the geometry that had been discovered by various mathematicians over the years. He systematically linked and logically ordered these geometric ideas.

There are many branches of mathematics that seem independent of one another, but if one were to look closely a connection would be evident. It is exciting to learn or discover these connections.

Consider the following ideas—

the Pascal triangle, Newton's binomial formula, the Fibonacci sequence, probability, the golden mean, the golden rectangle, the equiangular spiral, the golden triangle, the pentagram, limits, infinite series, Platonic solids, regular decagon —

all discovered by different people at different times in different places. Yet, these ideas have a thread that interconnects them.

The ***Pascal triangle*** is named after Blaise Pascal (1623-1662) although its earliest record appeared in a Chinese book printed in 1303. Each term of the Pascal triangle is the sum of the two numbers above that flank it. Each of its rows represents the coefficients of the binomial, $(a+b)^n$, raised to a particular power. The third row from the top row gives the coefficients for $(a+b)^3$. The nth row produces ***Newton's binomial formula.*** The ***Fibonacci numbers*** are found in the Pascal triangle by summing the numbers in the diagonals as shown.[1] The sequence is linked to many forms and phenomena in nature. There are many other sets of numbers in the Pascal triangle—look for the natural numbers, triangular numbers, tetrahedral numbers, 4-space tetrahedral numbers, 5-space tetrahedral numbers[2],... . ***Probability*** is linked to the Pascal triangle in various ways. The *normal distribution curve* is formed when a Pascal triangle of hexagonal blocks is made with balls falling from a reservoir atop the triangle. At each hexagon there is an equal chance of the ball falling to the right or the left. The balls collect at the bottom in the numbers of the Pascal triangle and form the bell shape of the normal distribution curve. Pierre Simon Laplace (1749-1827) defined probability of an event as the ratio of the number of ways the event can happen to the total possible number of events. The Pascal triangle can be used to calculate different combinations and the total possible combinations. For example, when four coins are tossed in the air the possible combinations of heads and tails are: 1 for four heads; 4 for three heads and one tail; 6 for

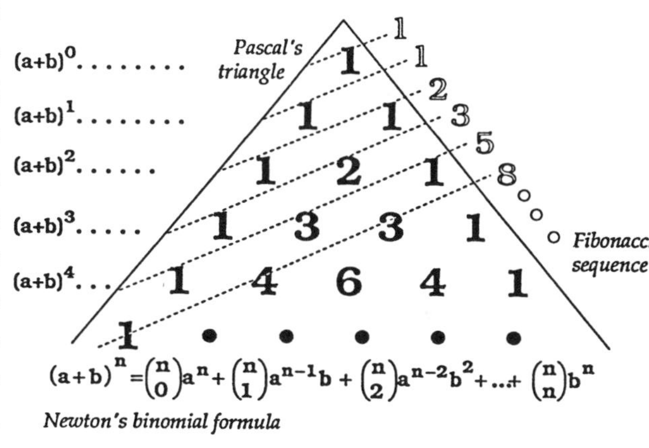

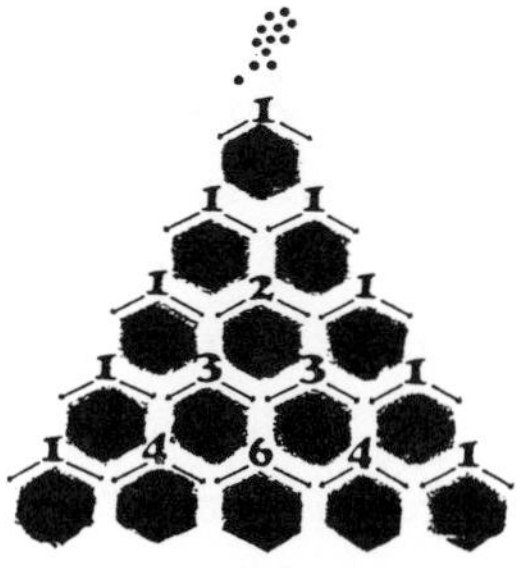

two heads and two tails; 4 for one head and three tails; and 1 for four tails. These numbers correspond to the third row after the top row of the Pascal triangle—1, 4, 6, 4, 1 —indicating the possible outcomes. Their sum is the total possible outcomes= 1+4+6+4+1=16. So to find the probability of tossing 3 heads and a tail, we get 4/16.

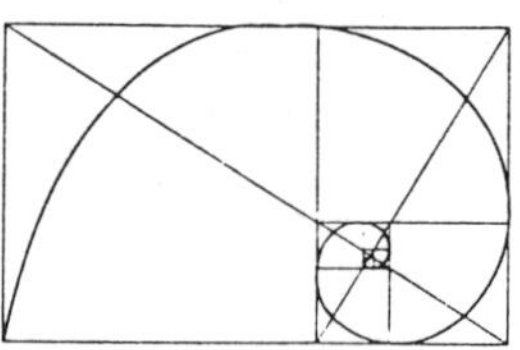
the golden rectangle & the equiangular spiral

The ***golden mean*** and the ***golden rectangle*** (used by the ancient Greeks in architecture and art) are linked to the Pascal triangle and probability via the Fibonacci sequence. When a new sequence of terms is formed from the ratio of consecutive terms of the Fibonacci sequence—1, 1, 2, 3, 5, 8, 13, 21, 34, ...— we get 1/1, 2/1, 3/2, 5/3, 8/5, ...,F_n/F_{n-1}. Each term of this sequence is slightly above or slightly below the golden mean, and in fact its limit is the golden mean, $(1+\sqrt{5})/2 \approx 1.618...$, the ratio of sides of the golden rectangle. The ***equiangular spiral*** is pulled into the picture by the golden rectangle. Since the golden rectangle is self-generating, starting with one golden rectangle infinitely others can be produced inside it as shown. An equiangular spiral can be formed from these golden rectangles. The intersection of the diagonals of the golden rectangle is the pole or center of the equiangular spiral. The golden mean connects us again to the ***golden triangle,*** an isosceles triangle with base angles 72° and vertex angle 36°, which is also self-generating and yields the equiangular spiral. Now the golden triangle gives a direct link to the ***pentagram,*** since the five points of

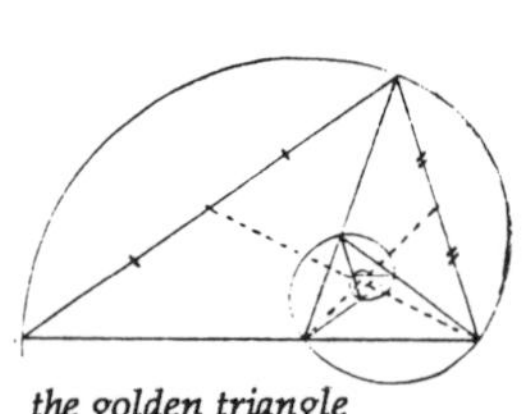
the golden triangle & the equiangular spiral

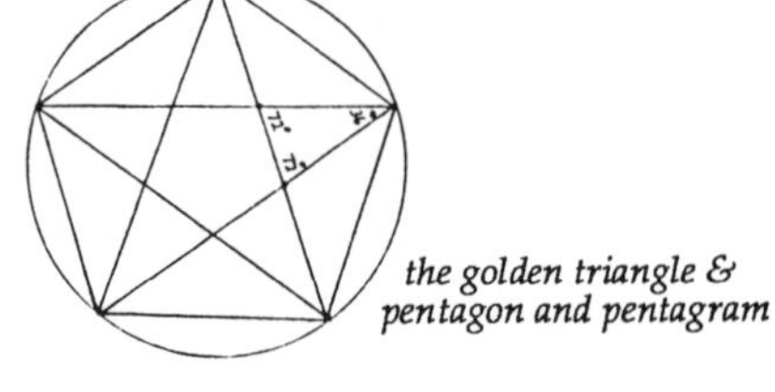
the golden triangle & pentagon and pentagram

the pentagram are golden triangles. We notice how ***limits*** come into the picture by looking at the limit of $F_n/F_{n-1} \rightarrow \varphi$ (the symbol for the golden mean). φ is also generated by other infinite series. Golden rectangles can also be used to draw the Platonic solids, ***icosahedron*** and ***dodecahedron..*** The icosahedron, 20-faced regular convex polyhedron, can be formed by three congruent golden rectangles that intersect each other symmetrically so that each is perpendicular to the other two, then their twelve vertices are the vertices of the icosahedron. The dodecahedron, a 12-faced regular convex polyhedron, is also formed from these three congruent golden rectangles. This time the twelve vertices are the centers of the faces of the dodecahedron. Lastly, the golden mean is linked to the ratio of the radius to the side of a regular convex ***decagon*** inscribed in a circle because the decagon can be broken up into ten golden triangles which have their vertices at the circle's center.

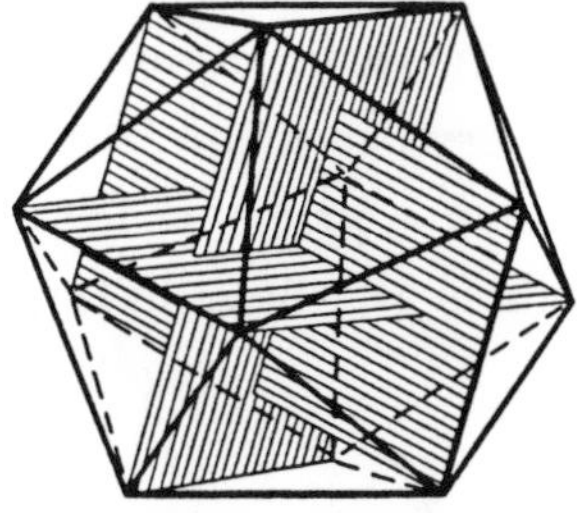

the golden rectangle & the regular icosahedron

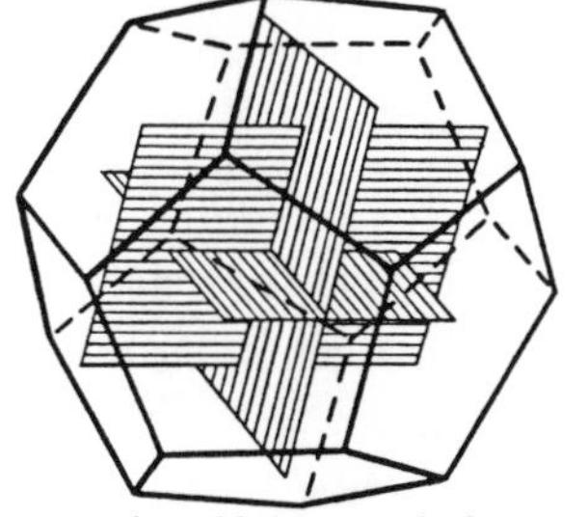

the golden rectangle & the regular dodecahedron

These connections evolved over thousands of years. As we see recognizing a common thread through mathematical ideas can be exciting and surprising.

[1]The Fibonacci numbers are a sequence of numbers that were discovered as the solution to a problem posed by Fibonacci (otherwise known as Leonardo da Pisa, 1175-1250) in his book *Liber Abaci* and reprinted in a book compiled in the 19th century by the French mathematician Edouard Lucas.

[2]See page 51 for a diagram of where these different numbers appear in the Pascal Δ.

prime properties

Prime numbers seem to hold a very special place in the realm of numbers. Every number has a unique prime factorization. For example, 12's prime factorization is 2•2•3 while 18's is 3•3•2.

Here are some of the unique properties of primes

1) *2 is the only even prime number*
2) *No prime number other than 5 can end in 5.*
3) *After the unit primes (2,3,5,7), all other must end in 1,3,7, or 9.*
4) *The product of two primes can never be a perfect square.*
5) *If a prime number other than 2 or 3 is increased or decreased by 1, one of the results is always divisible by 6.*

2, 3, 5, 7, 11, 13,
17, 19, 23, 29,
31, 37, 41, 43,
47, 53, 59, 61,...

Prime numbers have intrigued mathematicians for centuries. Christian Goldbach (1690-1764) wrote in a letter to Leonhard Euler that he believed it could be shown that every even integer other than 2 is the sum of two primes— e.g. 6=3+3; 8=5+3; 28=13+15.

What do you think?

Euler was unable to prove or disprove this. It still remains so today.

π is not a piece of cake

The π race has been going on for centuries, and there is still no winner. There is no finish line to reach. The lead changes from person to person with the most number of places in their estimate of π. The 530 million places reached by Gregory and David Chudnovsky seemed like a record that would have held for sometime. But no, Yasumasa Kaneda broke their record in August 1989 by carrying out π to 536,870, 000 places. This record filled 110,000 sheets of computer paper, and took 67 hours and 13 minutes on Japan's fastest supercomputer. Is there an end in sight? If the computer capabilities and capacity give out, then maybe π can be at rest.

3.14159265358979
32384626433832 79
50288419716939 93
75105820974944 59
23078164062862 08
99862803482534 21
17067982148086 51
32823066...

unusual paths of planets

It is almost startling to see these patterns and symmetrical designs describing the paths of planetary bodies. We often think that planets' paths are solely of an elliptical nature. The diagrams below show the paths traced by Mercury, Venus, Mars, Jupiter, and Saturn as seen from the earth.

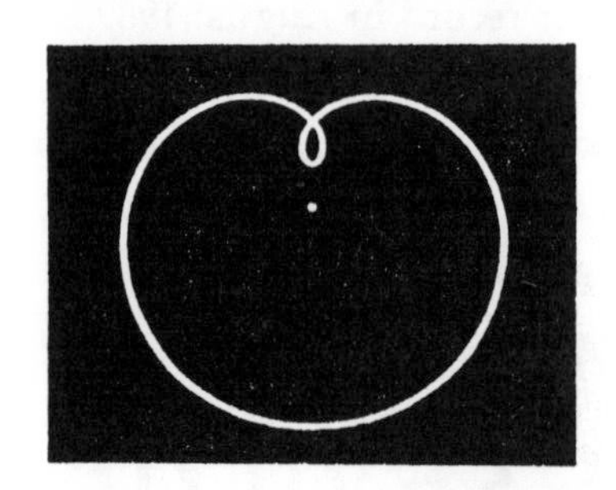

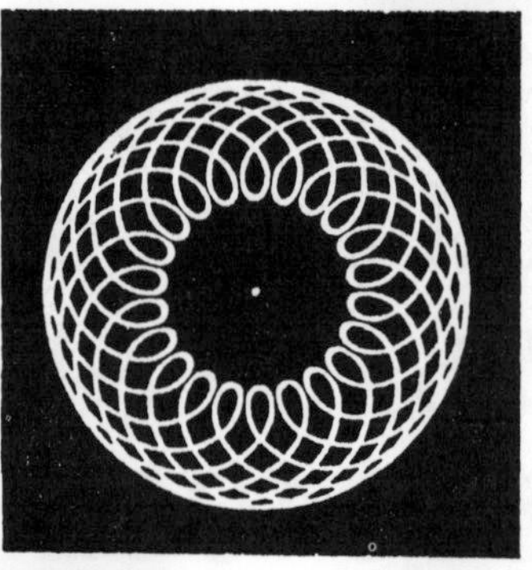

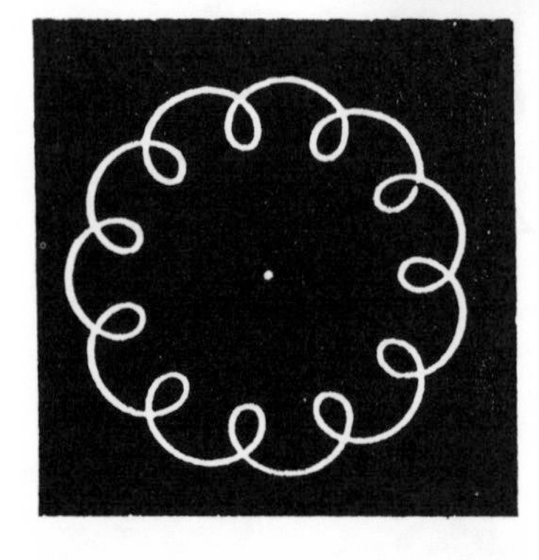

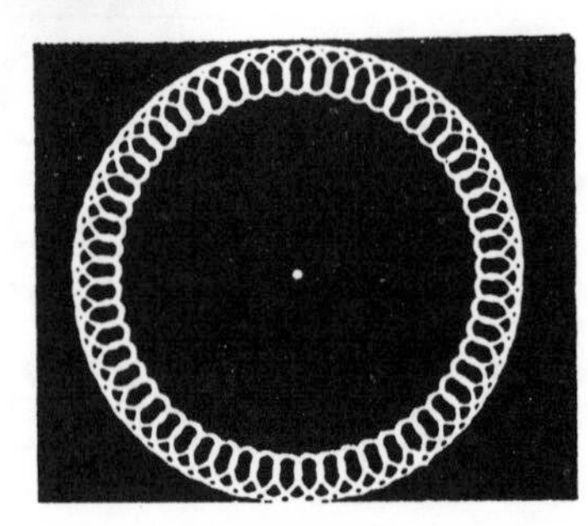

dice & the Gaussian curve

The normal distribution curve was first recognized by French mathematician Abraham de Moivre in the late 1600's. In the 19th century, Carl Friedrich Gauss further developed it by formulating the equation of the curve, and it has since borne his name.

The diagram below shows the 36 possible outcomes when a pair of dice are thrown. It is interesting to notice how the frequency of the outcomes of different possible numbers (2 through 12) illustrate the Gaussian curve.

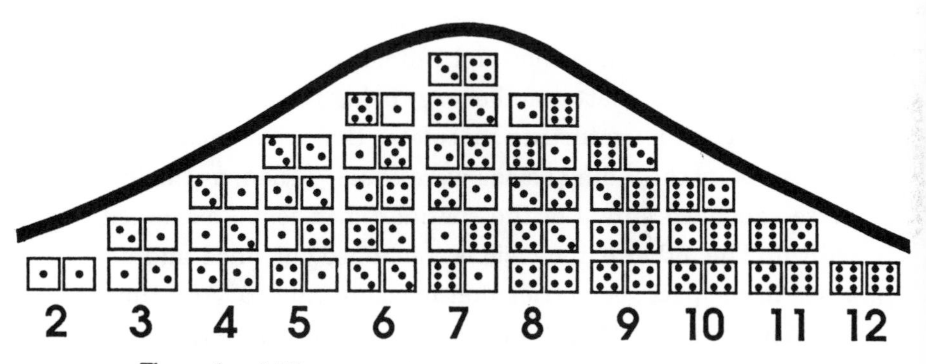

The number of different ways of tossing the numbers from 2 to 12 with a pair of dice.

mathematics' role in TOE– the theory of everything

How did the universe begin? Can such an overwhelming question be answered? In their incessant search of TOE (the theory of everything), scientists are trying to develop a mathematical model that would unify the forces of nature (electromagnetism, gravity, the strong force and the weak force). Physicists certainly believe a TOE can be found. Some have devoted their life's work in search of answers. In today's most current TOE, physicists believe that the essence of the universe is tied into objects called *superstrings*. A version of this particular TOE contends that the universe, all forms of matter and energy, from the instant of the Big Bang resulted from the actions and interactions of superstrings. The *superstring theory* describes the universe as 10-dimensional (9 spatial dimensions and 1 dimension of time) with its building blocks of matter and energy being these infinitesimal strings. The theory further speculates that at the moment of the Big Bang the 9-dimensions were equal. Then, as the universe expanded only 3 of the spatial dimensions expanded with it. The other 6 dimensions remained entwined and encased in compact geometries that measure only 10^{-33} centimeters across (i.e. 10^{33} laid end to end measure 1 centimeter). Hence, these strings are believed to possess 6 dimensions within them. Scientists are now trying to describe them by using 6-dimensional topological models. It is believed that these superstrings may either be open or closed in a loop, and make up the different forms of matter and energy of the universe by altering their vibrations and rotations. In other words, superstrings are distinguished among each other by how they vibrate and rotate.

The importance of the superstring theory is being compared to Einstein's Theory of General Relativity. Four dimensions were hard enough to envision. Einstein posed that length, width, height and time were needed to describe the location of

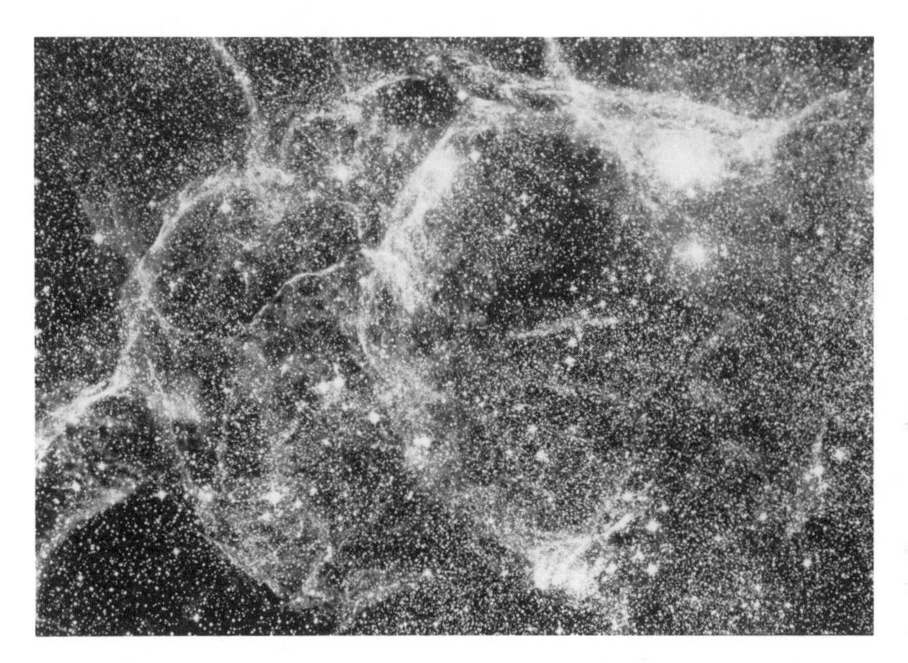

Mathematics plays a major role in the explanation of the universe and TOE.

an object in the universe. Ten dimensions seem out of the question. But if one thinks of dimensions as descriptive numbers that pinpoint the location of an object in the universe, they become comprehensible.

These ideas of TOE have been evolving for nearly 20 years. Gravity had been the wrench in the gears because the computations (involving gravity) needed to support the various forms of this theory produced mathematical infinities*. The breakthrough came in 1974, when John

Schartz and Joel Scherks considered gravity as a curvaceous piece of geometry in the 10th-dimension similar to how Einstein described gravity in the geometry of the 4th-dimension.

The mathematics behind the superstring TOE is very high powered. The results have been most convincing. Schwartz and Michael Green were two of the main initiators of this theory. They had been working on it for over a decade, in spite of little encouragement or support from colleagues, who found the 10-dimensional world hard to accept. But their published paper finally caused physicists to take the idea seriously. Some scientists contend physicists had been avoiding ("wasting" time) researching the idea because the mathematics is too difficult. Super symmetry, six-dimensional topological models, a ten-dimensional universe, infinitesimal strings are some of the concepts needed to describe and validate the theory. Consequently, this theory has turned some physicists into mathematicians and vice versa.

* Mathematical infinities can be caused by such operations as dividing a number by 0.

mathematics & cartography

We all know that the Earth is not flat, yet for portability and convenience we make and use maps drawn on rectangular sheets of paper. Since the Earth is spherelike, a map on a globe is the most accurate map. A global map shows — *all longitude lines as equal in length and meeting at the poles, all latitude lines as parallel, that latitude lines around the globe decrease in size as they approach the poles, the distances between longitude lines between any two latitude lines as equal, and the latitude and longitude lines meeting at right angles.* But it is not possible to draw an accurate map of the world on a flat sheet of paper. Consequently projections of global maps are made. Different types of projections make particular areas of a map more accurate. Concepts from projective geometry are utilized to create these different maps. For example, the *Mercator's projection* (cylinder or tube projection) is accurate for the area near the equator. Mercator's projection does not show lines of longitude converging at the poles. Thus, areas near the poles appear larger than they actually are. *Zenithal projection* on the other hand is used for accuracy in the polar regions. Here are other types of projections used in cartography — *azimuthal projections, conic projections, sinusoidal projections, homolographic projections, interrupted projections* — but if a projection is used, some portion of the map will be distorted. These distortions explain why navigators often use some maps or combinations of maps for various regions or types of navigation (aerial, ocean). Without knowledge of projective geometry, proportion, graphing, and spherical geometry cartography would be at primitive stages.

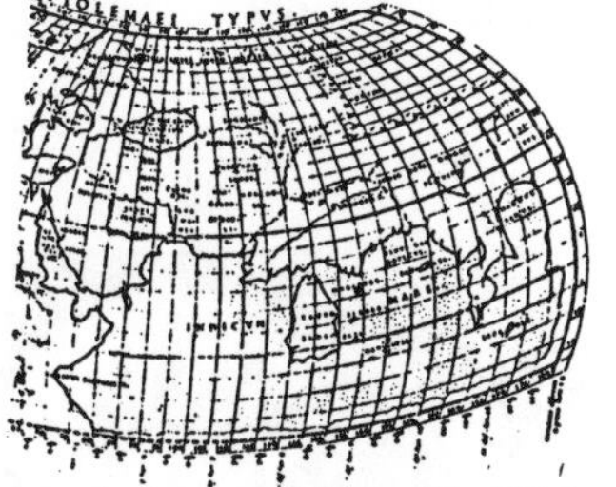

This is part of a map of the world of Ptolemy that was drawn by a Venetian cartographer in 1561.

spirals— mathematics in nature

Types of spirals are almost as numerous as the frequency with which they appear in all facets of life and nature. There are—flat spirals, three-dimensional spirals, right and left-handed spirals, equiangular, geometrical, logarithmic, rectangular spirals. When one thinks of curves, circles and ellipses come to mind most often. But there are a wealth of other curves which exist in mathematics and appear in nature and growth patterns of natural phenomena. Spirals fall into this category.

To get a feeling for a spiral, compare its properties to a circle. The distance around a circle, its circumference, is finite. The circle is a closed curve with all its points being the same distance from its center. On the other hand, the spiral has a starting point and winds around continually. Its length is infinite.

Since its starting and ending points are not joined, it is an open curve, and unlike the circle its points are not equally distant from the beginning point (its pole). Spirals can be two or three dimensional. A record is an excellent example of a planar (2-D) spiral. It is not made up of separate concentric circles, but is one single groove. Spatial (3-D) spirals, which are formed when spirals wind around different objects such as a cone or a cylinder— as in a DNA molecule, screw, or cork screw— are called helices.

the spiral in a fern's formation

Spirals are exciting curves to explore mathematically and discover their connection and formation in the growth of natural phenomena and other areas, such as —vines, shells, tornadoes, hurricanes, bone formations, vortices, the Milky Way, spider webs, architectural and artistic designs.

the remarkable equiangular spiral

The equiangular spiral is a fascinating curve. It appears in such growth forms of nature as the nautilus shell, a sunflower's seedhead, the webs of Orb spiders. René Descartes (1596-1650) was the first to study the equiangular spiral in 1638. In the late 1600s, John Bernoulli discovered many of its properties, and, in fact, was so taken with the curve that he asked to have it on his tombstone with the words—"Though changed, I rise unchanged."

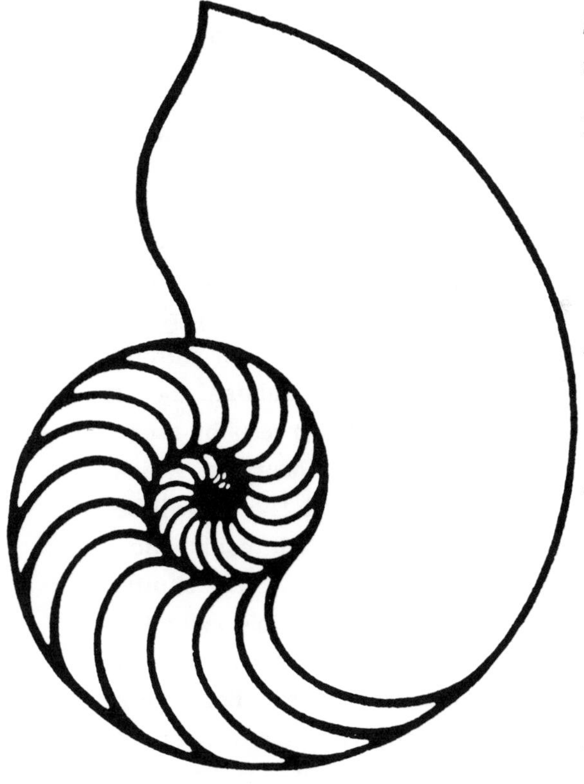

a chambered nautilus

Some of the properties of the equiangular spiral are:

1) *Angles formed from tangents to the spiral's radii are congruent. —hence, the term equiangular.*

2) $r=ae^{b\cot\theta}$ *—the polar coordinate equation for the equiangular spiral.*

3) *The equiangular spiral increases at a geometric rate, thereby any radius is cut by the spiral into sections that form a geometric progression.*

4) *While the equiangular spiral winds, its size changes while its shape remains the same.*

testing Einstein's General Theory of Relativity

Various parts of Einstein's theories have been proven over the years, but his General Theory of Relativity remains theoretical. In Einstein's Special Theory of Relativity, the constancy of the speed of light is one of its essential parts.

The speed of light, c, is 186,000 miles per second. Suppose you are in a spaceship traveling at half the speed of light, and are approaching a light source (a star). The speed of light from this star will always reach you at 186,000 miles per second. It is independent of your speed. The only way this can be explained is to theorize that perhaps the instruments in the spaceship were running slow, but nothing on your ship indicates this. Alerting the control station, you find their measurements show everything aboard your spaceship — animate or inanimate — has slowed down or shrunk..

The fact is that the measurements of space and time are dependent on speed—as speed increases time will slow down and distance (length or size) will decrease. This was illustrated in 1972, when two American scientists flew around the world in a jet equipped with an atomic clock. At the end of their journey the synchronized ground clock showed the clock aboard the jet had slowed by 89-billionths of a second. Other tests by linear accelerators have verified Einstein's Special Theory in which $E=mc^2$. In the near future, Einstein's General Theory will be tested. Stanford University scientists*, working in conjunction with NASA, are planning a space shuttle launch. The shuttle will carry special gyroscopes which will test Einstein's belief that gravity causes a curvature in space and time. Einstein claimed that massive bodies, as stars and planets, can curve both space and time. And it is this curvature that governs the movement of all bodies (small and large). Results in either direction will create a furor in the scientific world.

* Scientists at Stanford University have been working on this experiment for over thirty years.

creating triangles puzzle

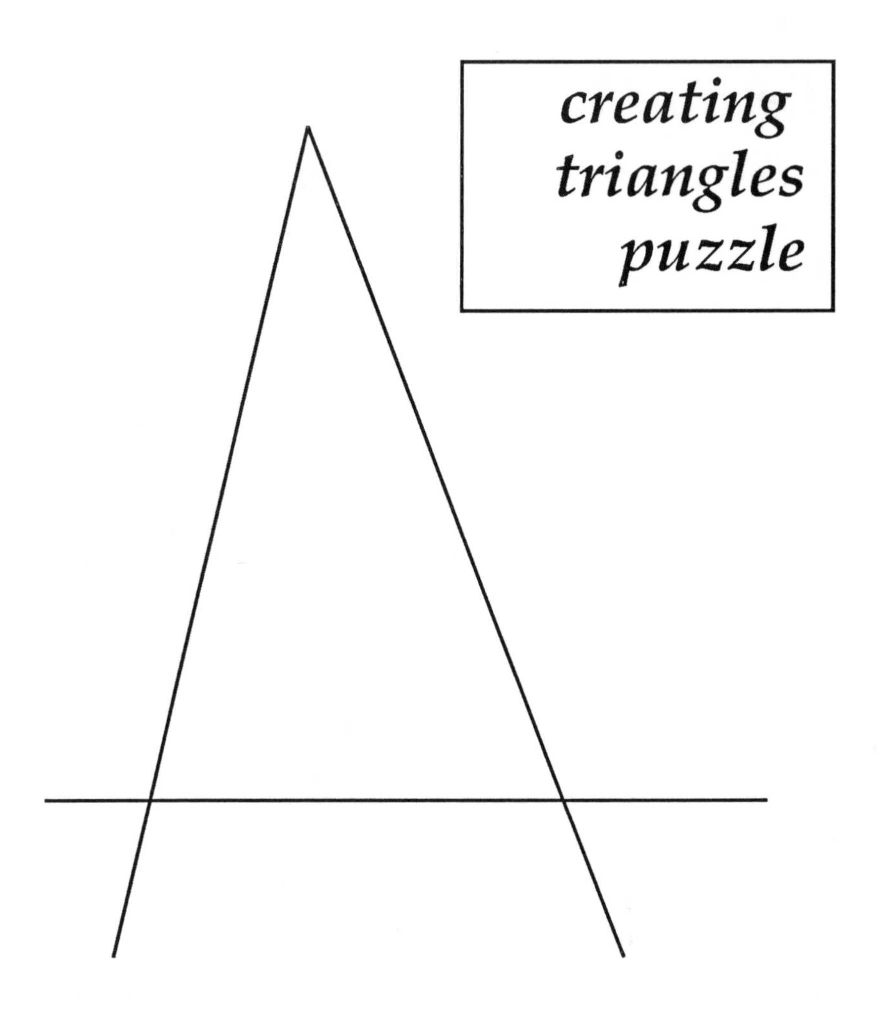

Add two straight lines to the above diagram and produce 10 triangles.

For solution, see appendix.

Fermat's last theorem— proven or unproven?

One of the most recent and famous unsolved *problems* of mathematics is *Fermat's Last Theorem.* Pierre de Fermat (1601-1665) was a lawyer by profession, who enjoyed spending his leisure time studying mathematics. In the margin of one his books he wrote—

> *To divide a cube into two cubes, a fourth power, or in general any power whatever above the second, into two powers of the same denomination, is impossible, and I have assuredly found an admirable proof of this, but the margin is too narrow to hold it.*

restated:

If n >2, there are no whole numbers a, b, c such that $a^n+b^n=c^n$.

If n>2, $\nexists$ whole numbers, a, b, and c.

$$a^n+b^n=c^n$$

Naturally, when this note was discovered after his death, the challenge was out to mathematicians. But for centuries the proof or disproof has eluded even the most prominent mathematicians. The theories that have been an outgrowth of efforts to prove Fermat's Last Theorem have become more significant than the theorem itself. Some feel he never had a proof, but intended to frustrate some of his colleagues. Nevertheless, it has stimulated an abundance of important mathematical ideas and discoveries over the past 350 years. In 1988, Japanese mathematician Yoichi Miyaoka presented his proof. Miyaoka retracted his proof less than six weeks after it was made public. He had tried an innovative idea in his proof, but obstacles arose which thus far seem to have no immediate solution. Consequently Fermat's Last Theorem is still around to challenge mathematical thought.

Möbius strip, π & Star Trek

Mathematics provides a wealth of ideas, such as the 4th dimension, Möbius strip, hyperspace, Klein bottle, π, from which writers of science fiction can draw. The concept of the Möbius strip was first created by German mathematician Augustus Möbius (1790-1868). In the episode, *Time Squared* in ***Star Trek– The Next Generation,*** the Möbius strip had a key role. In this episode the Möbius strip was applied to time. *The Enterprise* entered a "special time zone" that must have been shaped like the Möbius strip. They were trapped on this Möbius conveyor belt of time that would have endlessly repeated the same sequence of events, until the captain discovered a solution.

3.14159265358979323
84626433832795028841
971693993751058209
74944592307816406...

In another ***Star Trek*** episode, mathematics was the hero rather than the villain. Here π was used to foil a diabolic computer, when Spock asked the computer to give the number value of π. Naturally, since π is irrational (never ending non-repeating decimal) the computer was preoccupied until the crew could deactivate it. Hopefully other writers will use mathematical ideas in their creative writing.

Penrose tiles

Tile-like phenomena are evident everywhere—the plates of a turtle's shell, the scales of fish, and even the cells of our skin appear to fit together as if they were tiled. Over the centuries artists have used tiles to tessellate mosaic floors, paintings and walls. Moslem artists became masters of geometric designs in tessellations. M.C. Escher animated and expanded their work by tiling with objects that appeared to be in motion, such as birds, people, animals, fish. All these forms of tessellation are called *regular periodic tilings.* In periodic tiling, a basic pattern in the design can be seen to repeat on a regular basis, when the eye moves vertically or horizontally.

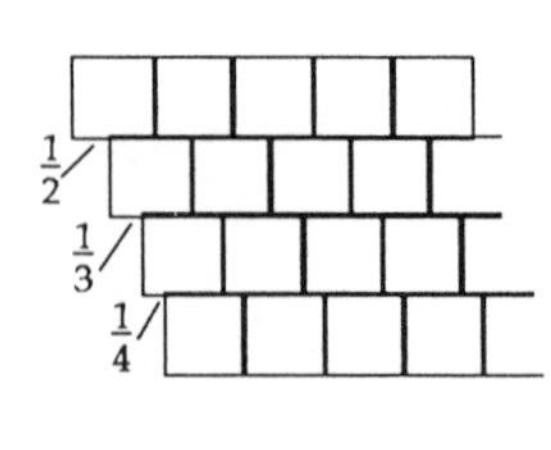

The example on the left is of a *non-periodic tiling* made using square tiles placed in rows which are staggered. On the right is an example of non-periodic tiling using right triangles.

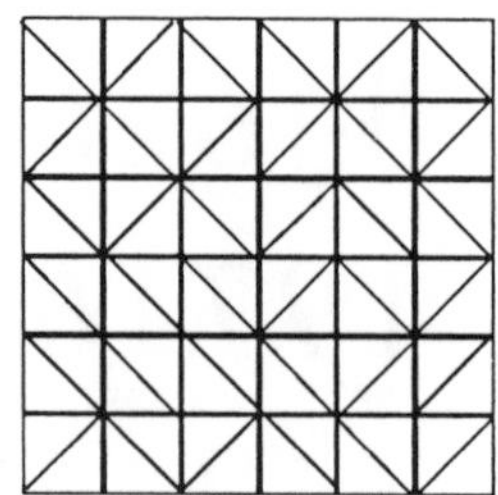

Mathematicians believed that if a non-periodic tiling could be made with particular shapes, then a periodic one could also be made with the same shapes.[1] In 1964, a set of tiles was discovered that *only* allowed a non-periodic tiling. This set consisted of 20,000 different shaped tiles. With this discovery others followed that required far fewer tiles. In 1974, Roger Penrose[2], the British mathematical physicist, discovered a set consisting of *only* two tiles that could produce an infinite number of different non-periodic tilings of a plane. He named his tiles

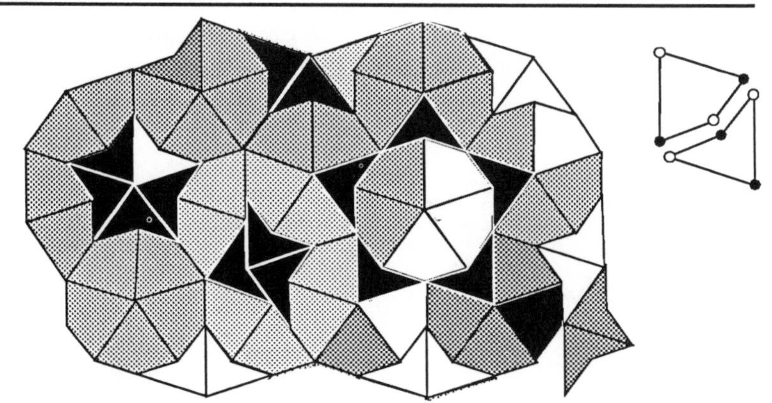

darts and *kites*. They are constructed as illustrated using a rhombus[3], the golden mean and the golden triangle.

The darts and kites must be joined at vertices with matching circles, as shown. The diagram below illustrates the seven kinds of vertex clusters that occur in all Penrose tilings of darts and kites.

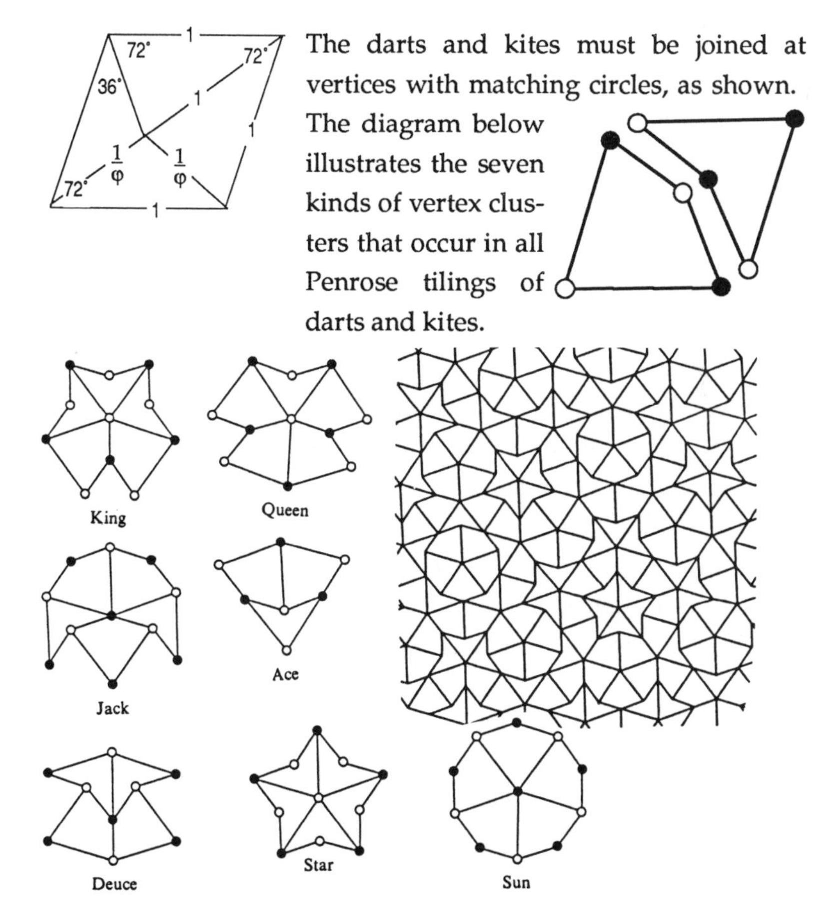

In addition to nonperiodic tilings with darts and kites, the tiling shown here is formed from rhombi that can be developed from the darts and kites as follows.

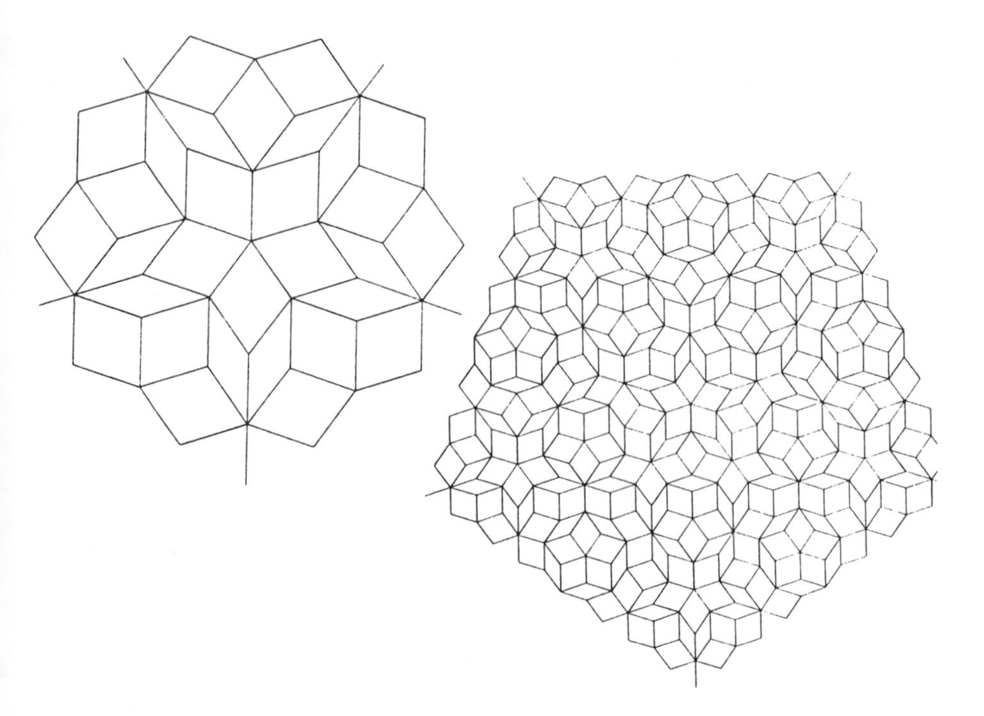

Although Penrose tiles do not form periodic tiling, they possess a type of symmetry, called fivefold and tenfold symmetry. For example, a Penrose tiling copied onto a transparency and rotated 1/5 around will coincide with an existing pattern.

All too often mathematical ideas do not seem to stimulate interest outside of mathematical circles. Penrose's exciting discovery was no exception. But in 1982 a practical application was discovered that made the non-mathematical world stop and take notice. Chemist Daniel Shechtman, discovered a way to join manganese and aluminum to make a new super strong alloy. But the crystals he discovered appeared to defy an existing law, Barlow's law[4]. The crystals of this alloy possessed fivefold symmetry. As a result many scientists did not take Shechtman seriously because they had always felt crystals could not have fivefold symmetry. Until this new alloy, all rotations around a crystal's axes produced identical patterns if rotated 1/2 turn, 1/3, 1/4, 1/6—but 1/5 turn was believed to be mathematically impossible.[5] When scientists began to realize a connection with Penrose 2-D tiles, they utilized his idea and created 3-D models of Penrose tilings to describe certain crystal formations.

Penrose tiles stimulated many ideas, fascinating mathematical implications, and "new worlds". These two simple shapes act as the bases for creating an infinite number of different tessellations. Discover properties and the various ways the darts and kite must fit together by creating your own Penrose tilings.

Mathematicians have found many other sets of tiles which create non-periodic tilings. But an unanswered question remains—Is there a single-shaped tile that only tiles a plane non-periodically?

---

[1]The examples of non-periodic tilings made with staggered squares and right triangles can be arranged into periodic tilings.

[2]The *impossible tribar*, used by such artists as M. C. Escher, was created by Roger Penrose. He is also responsible for the theory of twistors. Although twistors are not visible, he believes that space and time are interwoven by the interaction of twistors.

[3]A rhombus tiles periodically. Rhombi (made from of a darts & kites) placed side by side would create a periodic tiling. Thus a Penrose tiling would avoid the use of rhombi placed in this way.

[4]Barlow's law, a mathematical theorem, states that 5-fold symmetry is impossible in periodic tiling of a plane and space. The scientists using this theorem in their work apparently had not considered non-periodic tiling of crystals possible.

[5]Apparently the argument is analogous as to why congruent regular pentagons cannot tessellate a plane.

place value number system —Where did it come from?

Today we take for granted the concept of place value. We are taught to write numbers in base ten, and it becomes second nature to us and seems the natural way to count. It is astonishing to realize that about 28,000 years lapsed from the first time a mark or stick was used to signify an amount to when the notion of place value was conceived by the Babylonians around the second millennium (2000 B.C. to 1000 B.C.). It is equally surprising that the first place value system was not base ten but a type of base 60. During this span of 28,000 years many different symbols for numerical quantities were developed by many cultures. Without a place value system to represent a quantity it was necessary to repeat symbols until the amount of the quantity was reached. For example, the Egyptian numerals were originally hieroglyphic script, primarily based on the repetition principle. To write a quantity they would repeat their symbols: 𓏺 =1 , 𓎆 =10, 𓍢 =100, 𓆼 =1000 and so forth. To write the symbol for 34 in hieroglyphic Egyptian numerals one writes 𓎆𓎆𓎆𓏺𓏺𓏺𓏺 . The oldest known place value system is the Babylonian, which was based on the Sumerian sexagesimal numbers. But instead of needing sixty symbols to write the numerals from 0 to 59, two symbols , 1= 𒁹 and 10= 𒌋 , sufficed. One major difficulty with the system was the initial absence of zero. (Imagine if our system did not possess a symbol for zero; 202 would be confused with 22, 2002, 220 etc.) If a number required a placeholder its value had to be determined from the context of the tablet. Eventually a space was left as a placeholder by scribes. This was an improvement, but did not solve the problem of numerals which ended in zeros, and again their values had to be determined from the context of the tablet. Sometime between the 4th and 1st century B.C., a symbol for the zero placeholder was developed, or . With the invention of zero, the Babylonians were now able to write numerical expressions for fractions. A number starting with the zero symbol indicated that it was a fraction, for example = 0°15' 20" (=0+15/60 +20/360). The notion of using the symbol to represent *nothing* , a zero quantity, was not immediately conceived or connected to the zero symbol.

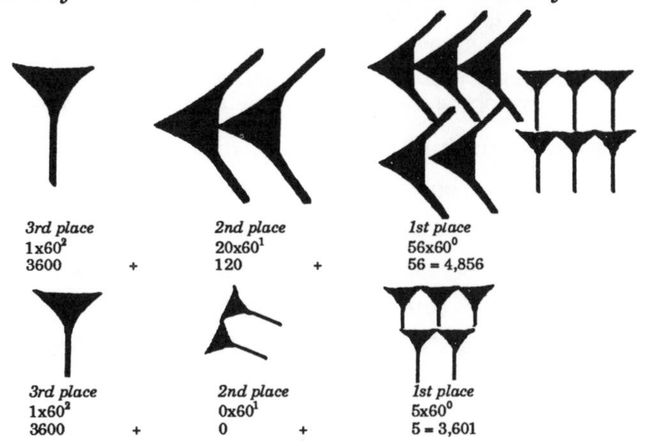

Sometime during the first half of the second millenium B.C., Babylonian scholars developed the first true place value numbering system.

This idea evolved with the passage of time. First Greek and later Arab and Jewish astronomers adopted the Babylonian notation for their astronomical tables by adapting their alphabetic numerals in place of the cuneiform numbers. It was not until 500 A.D. that a Hindu invented a positional notation for base 10 system. The alphabet letters past nine were thus abandoned, the symbols standardized and place value incorporated.

Today we still find evidence of the Babylonian sexigesimal place value system. Sexigesimal units are used in measuring angles (degrees, minutes, seconds) and time (hours, minutes, seconds). If base value systems had not been invented, can you imagine the horrendous task of computation using a system similar to the Roman numerals? And how could computers have been developed or programmed if it were not for base two? If the selection of a base for place value did not

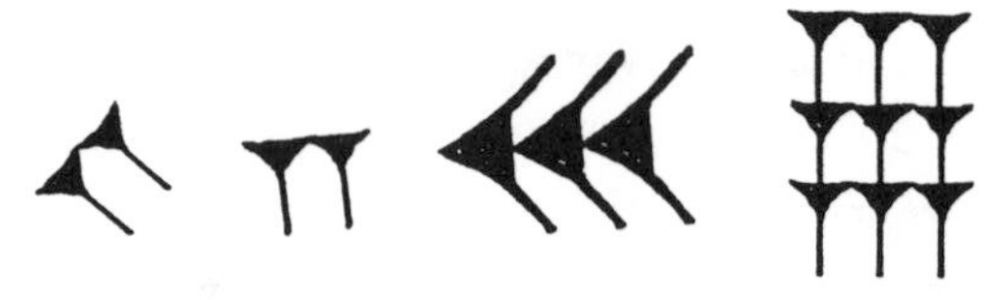

one's pl. 60th pl. 3600th pl. 216000th pl.
0x60^0 2x60^{-1} 30x60^{-2} 9x60^{-3} = 0 + 2/60 + 30/36,000 + 9/216,000

Babylonian astronomers were able to indicate sexagesimal fractions by starting the number with the zero symbol in the one's place.

involve a gradual evolution of ideas, but was left to a group decision, the base ten may not have been the one selected. For example, in the 18th century, Comte George Louis Leclerc de Buffon proposed base 12 be universally adopted. 12 has four divisors compared to base 10's two, but the famous mathematician Joseph Louis Lagrange advocated a prime base because every systematic fraction would be in reduced form and therefore represent a unique number. For example, in base 10 the decimal .40 represents at least three possible fractions,

40/100 =4/10 =2/5.

When a fraction in a prime base is converted from its decimal form, its denominator is solely composed of factors of that prime number. Thus the prime base would cut down on the possible fractions that are equivalent to it. Examples— .24 = 24/7^2cannot be reduced and has no equivalent fraction; .020=20/7^3, which reduced to 2/7^2. One can't help but wonder if perhaps it was our ten fingers that gave the final vote for base 10.

On which day were you born?

We can easily determine the day of the week a certain event took or will take place.

MONTH NUMBERS	
January	1
February	4
March	4
April	0
May	2
June	5
July	0
August	3
September	6
October	1
November	4
December	6

If it's a leap year, January is 0 and February is 3.

DAY NUMBERS

Sunday is 1
Monday is 2
Tuesday is 3
Wednesday is 4
Thursday is 5
Friday is 6
Saturday is 7

CENTURY NUMBERS

For date from– Sept. 16, 1752 to 1700 add **4.**
For date from–1800 to 1899 add **2.**
For date from–1900 to 1999 add **0.**
For date from–2000 to 2099 add **6.**
For date from–2100 to 2199 add **4.**

Example: to test August 27, 1990.

steps

1) *Take the last two figures in the year—for 1990, we use 90.*
2) *Divide this by 4 and ignore its remainder, if any. 90÷4=22 (and remainder 2 you ignore).*
3) *Go to the month chart and find the number for August. It is 3.*
4) *The day of the month for our date is 27. Add it to the numbers we got from steps 1 through 3. We get 27+90+22+3=142.*
5) *Divide this sum by 7. We get 142÷7=20 plus remainder* **2.** *(Note: if the remainder is 0 use 7 for the remainder number).*
6) *From the century chart, find the century number for 1990. We see it's* **0.** *Add this to the remainder from step 5.*

The remainder + the century number = 2+0=2.

The sum, **2**, tells us the day of the week. Find the number in the days numbers chart. We see 2 is for Monday.

On which day were you born? How does it work?

the projection of a hypercube

Plane projective geometry studies the properties of planar objects that remain unchanged when an object is projected onto another plane. For example, when figures of one plane are projected through a point to another plane certain characteristics of those figures remain while others change. Lines project into lines, triangles into triangles, but circles may not necessarily project into circles. Depending on the point of projection, they may become ellipses.

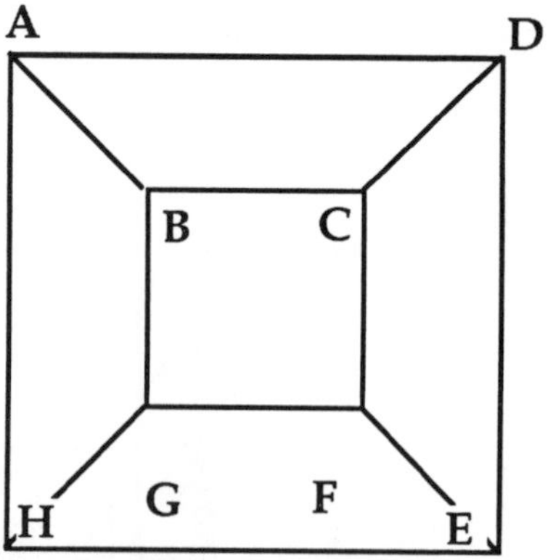

A D B C H G F E

The six squares are ABCD, ABGH, HGFE, CDEF, ADEH, and BCFG.

In all cases, distance, angle measurement, and congruency are not preserved.

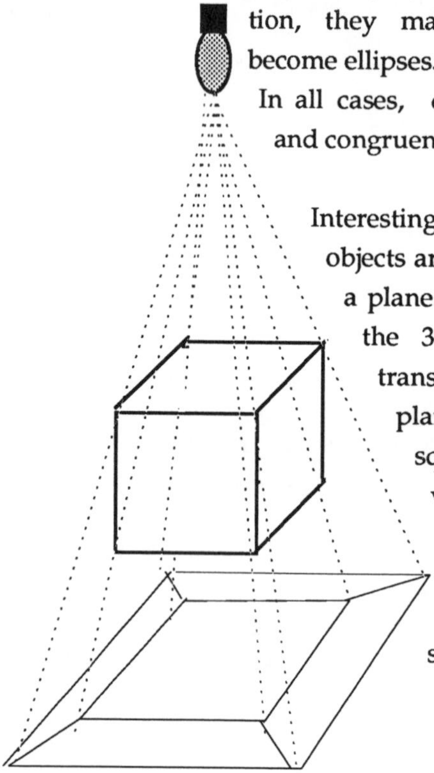

Interesting properties of 3-dimensional objects are revealed when projected on a plane. The diagram below shows the 3-dimensional cube's images transferred onto the 2-dimensional plane. The result — the six square faces of the cube can be viewed at the same time. To really tax your imagination, can you envision the projection of a hypercube onto space?

the Einstein "cover-up"?

The February 1990 meeting of the American Association for the Advancement of Science sparked heated discussion on whether Einstein should receive full credit for his theories. Senta Troemel-Ploetz's book, *In the Shadow of Albert Einstein: The Tragic Life of Mileva Einstein-Maric* is the cause of the furor. Einstein and Maric met while attending the Swiss Federal Institute of Technology in 1896. They were married in 1903, and divorced in 1914. The author contends that Einstein's work was a cooperative effort, citing the following—

$$E=mc^2$$

1) They took almost identical courses.

2) They wrote their final theses in the same area.

3) Both failed their final exams in 1900, yet Einstein was allowed to graduate.

4) Until 1982 Swiss universities were allowed to have different academic standards for men and women.

5) Maric would not have been allowed to study physics at this predominantly male school if she had not been brilliant.

6) Maric reportedly told her father that she and Einstein had recently finished a very important work that would make Einstein famous. The same year, Einstein published his paper on Special Relativity, Brownian motion, and photoelectric effect (for which Einstein was awarded the 1921 Nobel Prize).

7) Einstein gave Maric his entire Nobel Prize money, which she used to pay for the care of their psychotic son.

8) Continual references are made in the 41 letters from Einstein to Maric to "our research" and "our work". Only 10 letters of Maric to Einstein remain. None of them mention physics. Yet the 41 letters from Einstein to Maric always have reference to physics, and Einstein often acknowledges points of physics and theory that Maric had raised in letters not extant.

9) Maric worked with physicist Paul Habricht on the development of a machine to measure small electric currents, which was patented under the name Einstein-Habricht.

10) The author points out that at that time it was common for men to appropriate women's work and take credit for it. In addition, the couple may have agreed to keep Maric's contribution secret to enhance Einstein's chance for a university appointment.

the mathematical shuffle

With a trained hand, the cards of a deck can be made to perform "magic". For example, a 52-card deck can be returned to its original order with 8 perfect shuffles. A perfect shuffle requires layering the halves of the deck. But when mathematical magicians such as Persi Diaconis of Stanford University get into the act, a formula appears that reveals all the ways any size deck can be ordered by performing repeated perfect shuffles[1]. Unfortunately, this feat is not easy.

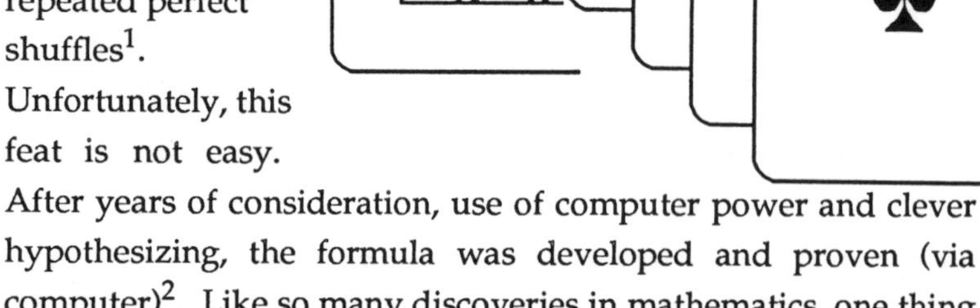

After years of consideration, use of computer power and clever hypothesizing, the formula was developed and proven (via computer)[2]. Like so many discoveries in mathematics, one thing leads to another mathematical connection — in this case the possible arrangement for each size deck turned out to be linked to group theory.

1 There are two types of perfect shuffles—one where the top card remains on top and one where it becomes the second card from the top.

2 Persi Diaconis and Ronald Graham from Bell Laboratories and William Kantor from the University of Oregon worked cooperatively to discover the ordering formula.

mathematics & superstitions

People over the ages have given special significance to numbers. Some people believe certain numbers are lucky or possess other powers besides simply describing a particular quantity. Many people have a lucky number, and people from various cultures find numbers such as 13 to be unlucky.

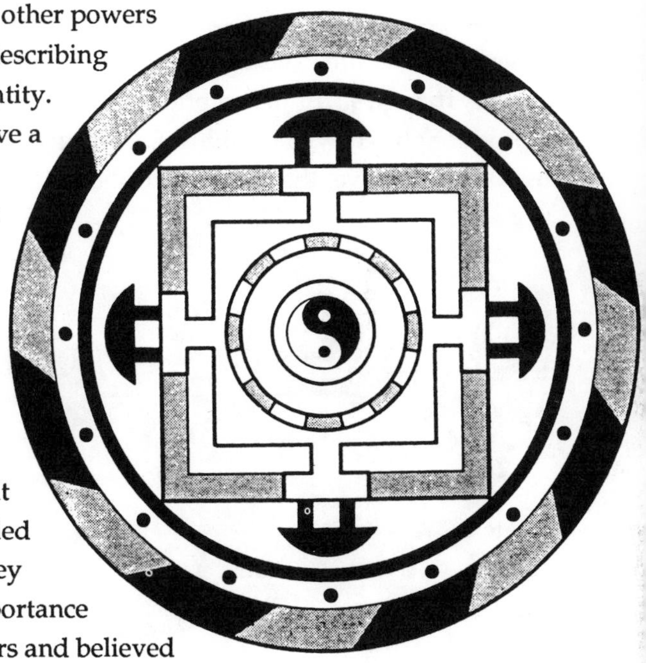

The Pythagoreans felt that numbers ruled the universe. They gave special importance to whole numbers and believed if you were able to master their use you could understand and influence the course of the universe. They even felt numbers affected reason, health, justice, marriage. For example—

The mandala was an ancient religious symbol in which the number 4 was its essence — usually a circle with a square or divided into quarters or multiples of 4.

1 was the origin of all numbers. Even numbers were female, and 2, the first even number stood for diversity of opinion. 3 was the first male number and was considered harmonious since it was composed of 1 and 2, unity and diversity. 4, being a perfect square represented justice. 5 represented marriage, since it was composed of the first female and male numbers.

mathematics, fractals & dragons

Fractals have come to be referred to as the geometry of nature. Though there are abundant examples of objects from Euclidean geometry present in nature (hexagons, circles, cubes, tetrahedrons, squares, triangles...), the so often randomness of nature seems to produce objects which elude a Euclidean description. In these cases fractals give the best description. We know Euclidean

geometry is great for describing such objects as crystals and bee hives, but one is hard pressed to find objects in Euclidean geometry to describe popcorn, baked goods, bark on a tree, clouds, ginger root, and coastlines. While Euclidean geometry has its roots in ancient Greece (circa 300 B.C. Euclid wrote the *Elements*), fractals date to the late 1800's. In fact, the term fractal was not coined until 1975 by Benoit Mandelbrot.

There are two categories of fractals, *geometric* and *random*. The properties of fractals are varied. For example, the dimensions of fractals on a plane are fractional, that is a dimensional number between 1 and 2, while fractals in space have a fractional number dimension between 2 and 3. In the world of fractals we can no longer say it's 2 or 3 dimensional, rather some may be 1.75-dimensional and others 2.3-dimensional. A coastline, in fractal geometry, is considered to be infinite in length because every little inlet and unit of sand is measured and the number of these units is constantly changing, just as in the formation of the dragon curve.

There are many forms and uses of fractals. This group of fractals has the property that *their detail is not lost as it is magnified* — in fact the structure is identical to the original fractal . An example is this illustration of the Cesaro curve.

Cesaro curve

New applications for fractals are constantly being discovered. Since fractals can be described by recursive functions (e.g. the Fibonacci sequence is a recursive sequence since it can be generated by adding its two previous terms) computer programs are ideal for generating fractals. Computers have been used to generate fractal landscapes and scenes in such movies as *Star Trek II: The Wrath of Khan* in the birth of a new planet and the planet floating in space in *The Return of the Jedi,* as well as the superb graphics displayed by Pixar Company at a Dallas computer show in 1986. Fractals are even being used to describe and predict the evolution of different ecosystems (e.g. Okefenokee Swamp in Georgia[1]). In fact, ecosystem work with fractals is now essential in determining the spread of acid rain and other pollutants in the environment.

Fractals have certainly opened up an entirely new and exciting geometry. This very new field of mathematics touches many areas of our lives – the description of natural phenomena, cinematography, astronomy, economics, meteorology, ecology (to name a few) — and generates rather bizarre objects with far out properties. Fractals are especially interesting because their applications are so extensive and their characteristics so fascinating. Here we have the geometry that describes the ever changing universe.

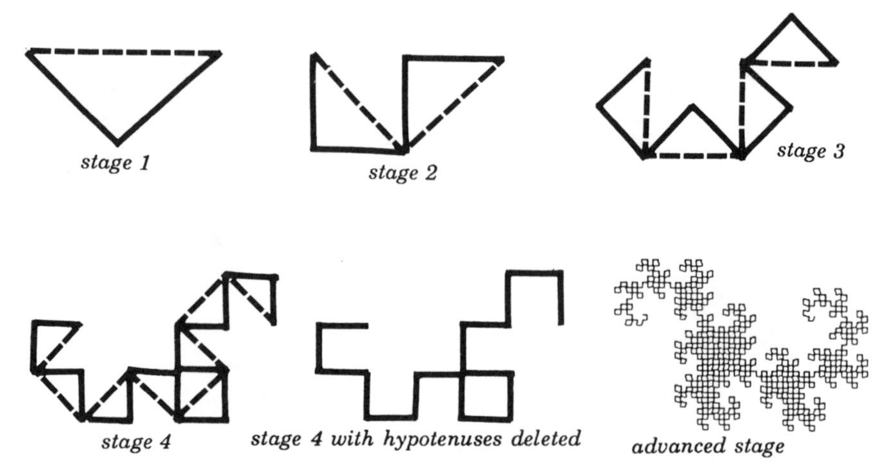

Constructing a dragon curve using isosceles right triangles. Dotted lines indicate the hypotenuses to be deleted.

The dragon curve was originally discovered by physicist, John E. Heighway. There are a number of ways to generate it. The method used here is the same used to generate the snowflake curve. In the snowflake curve we start with an equilateral triangle and continually add smaller equilateral triangles off the trisected sides. With this dragon curve start with a hypotenuseless isosceles right triangle. Off opposite sides of the legs make another hypotenuseless isosceles right triangle by considering the leg as the new hypotenuse and remembering to delete it. Now try creating your own fractal starting with some other type of geometric object and devising a similar procedure.

[1]Harold Hasting, a mathematician at Hofstra University N.Y. used fractals to model ecosystem dynamics at Okefenokee Swamp. Vegetation and cypress patch maps were compared with random fractal maps. Results demonstrated how species survive in competition with other species without need of extensive past histories.

overlapping squares problem

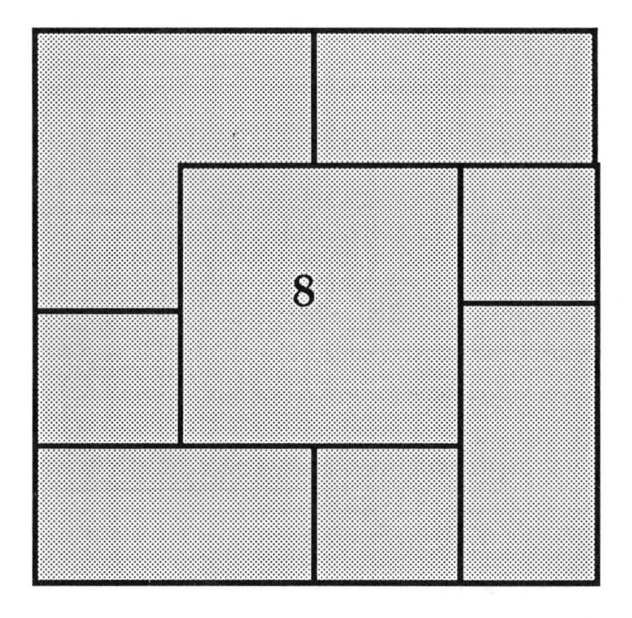

Eight congruent squares are layered one on top of the other. If the square numbered 8 was the last to be placed, determine the order in which the other 7 squares were placed to end up with the arrangement pictured.

For solution, see appendix.

the exponential power of Japanese swords

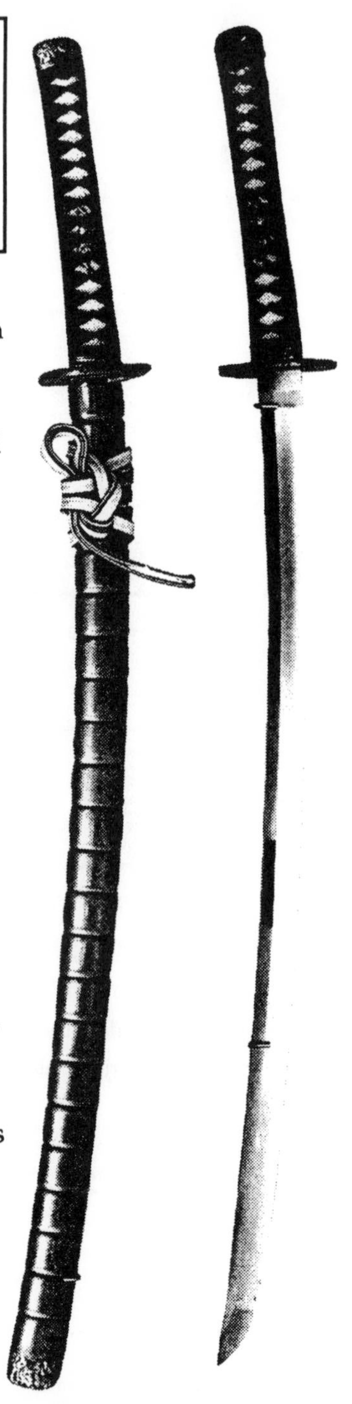

The art of sword making in Japan was an old and revered profession which was secretly passed from father to son or master to pupil. The swordsmith followed specific religious observances and traditions, and wore a ceremonial costume.

First, the swordsmith had to take special care and skill in welding a strip of steel to a rod of iron, which served as the handle. Other steel strips were placed on it and welded into a bar of specific dimensions, e.g. 6-8" in length and 1.25-2" in width. Since the sword had to be both flexible and hard, it had to be constructed in layers. The bar temperature was raised to welding heat. It was then folded, welded, and forged to its original dimensions. To avoid contact of grease and possible flaws, care was taken that the metal never touched the hand. The process of doubling it over, welding, and forging was done repeatedly. In fact, it was done 22 times, which produced 2^{22}=4,194,302 layers of steel! Between each forging the bar was cooled in oil and water alternately, then drawn out by hammer to the desired length and shape.

Anti-snowflake curve

Let's discover what happens when we reverse the process of forming the snowflake curve*.

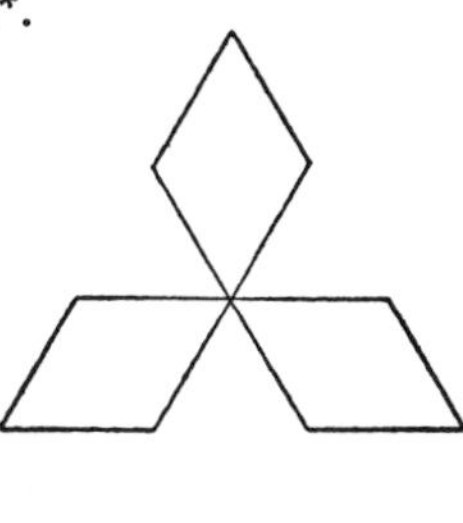

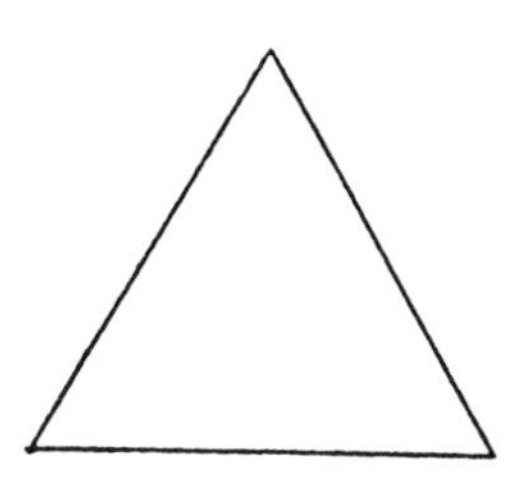

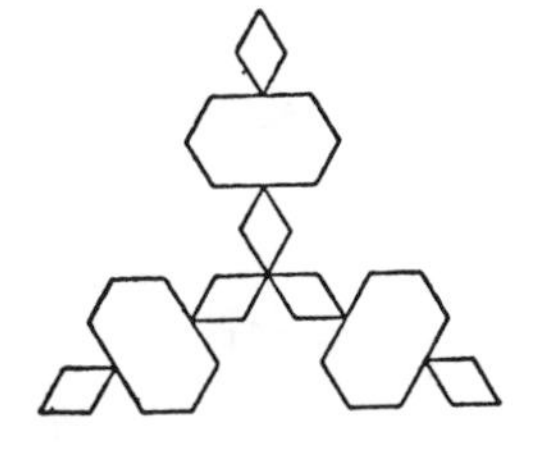

To generate a snowflake curve, we begin with an equilateral triangle. Next trisect each side of the triangle. Now on each middle third make an equilateral triangle pointing outward, but delete the base of each new triangle that lies on the old triangle. Continue this procedure with each equilateral triangular point—trisecting the sides and drawing new points. The snowflake curve is generated by continually repeating this process.

If we draw the triangles inward rather than outward we generate what is called the anti-snowflake curve.

Like the snowflake curve, the anti-snowflake curve has an infinite perimeter, but a finite area. This fact allows one to be able to draw the curve on a sheet of paper without running out of space.

*The snowflake curve is a geometric fractal. It was first used by Helge von Koch in 1904 in his study of curves with infinite length that contain a finite area. It is also called the Koch curve.

Mixing mathematics & baseball —high tech baseball

In 1845, The New York Herald ran the first box baseball statistics. These early reports measured the runs scored and outs totaled by each player. This early analysis is the forerunner of today's computerized statistical baseball data. Today we find that the data includes a range of information from runs, walks, balks, sacrifice flies. There are many variables to consider. For example, if runs are used as a basis for statistics, how are homeruns distinguished from stealing home? What about players on base that are brought in? For base hits, how are single base hits distinguished from a triple? What is done with pitching and fielding information? How about a player's ability to steal bases versus his batting ability? The computer and linear programming can bring in many valuable factors. Stir in statistical analysis, and what do you have?—sabermetrics*. Sabermetrics studies the entire realm of data available in attempts to give the best analysis of the information. In fact, in the book *The Hidden Game of Baseball* by John Thorn and Pete Palmer we find a formula analyzing offensive work in runs produced.

Runs=
.46(singles) + .8(doubles) + 1.02(triples) + 1.4(home runs)
+.33(walks+hit-by-pitches) +.3(stolen bases) -.6(caught stealing)
- .25(at bats–hits) -.5(outs on base)

Naturally, to have this data available on all players at a game and to quickly perform the calculations, would require a computer in the dug-out. Is this too far-fetched? Let's play ball.

---

*Sabermetrics stands for Society of American Baseball Research, founded in 1971 to promote the historical and statistical study of baseball.

Cretan Numerals

One of the earliest European civilizations was on the island of Crete from 2200 to 1400 B. C. The Minoan hieroglyphic writing, Linear A, has still not been deciphered, but another form of writing they evolved, now called Linear B, which uses schematic symbols rather than glyphs was deciphered after World War II by A. Ventrist. In archaeological digs, clay tablets and bars along with vases, religious tablets, stamps, labels and copper ingots show examples of Linear A and B. The clay tablets emphasize an accounting system the Minoans used to take inventories, and the clay bars probably functioned as accounting and data storage devises. The number system was a decimal system which relied on the repetition of symbols, rather than place value, to write a number.

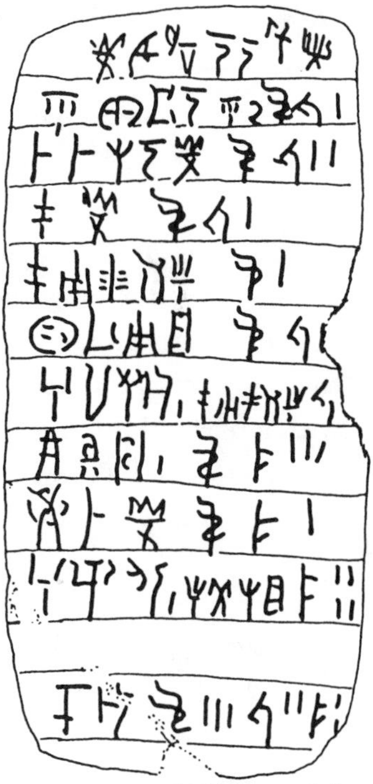

A tablet excavated from Knossos.

They used

I or) for 1

● or O for 10

/ or \ for 100 and in Linear A ◈ was used for 1000

To write 160 OO OO OO /

In addition, excavations have revealed tablets in which writing was in both directions (right to left and left to right).

Ada Byron Lovelace & computer programming

The Analytical Engine has no pretensions whatever to originate anything. It can do whatever we know how to order it to perform. It can follow analysis, but it has no power of anticipating any analytical revelations or truths. Its province is to assist us in making available what we are already acquainted with.
–Ada Bryon Lovelace

The illustration on the following page shows an array of some of the computer languages that have been devised over the years. There probably are as many computer languages as there are foreign languages. When a programmer is working with a particular language that does not suit his or her needs, it's altered and a new computer language is born. But when, where and how did these originate?

The story begins with the designs of the difference and analytical engines – the first computers – by Charles Babbage (1792-1871). Unfortunately, the technology of the times was not adequate to sustain his computers which could be *programmed* to make computations, take *input,* and follow the instructions of the program.

Although her name is not usually found in historical math books, Ada Byron Lovelace goes down in history as one of the first computer programmers. Born in 1821 to Augusta Byron and the English poet Lord Byron, Ada always had a keen interest and passion for mathematics. Living at a time when women were often discouraged and excluded from studying sciences, she was fortunate to have Mary Sommerville as a family friend of whom to ask mathematical questions and

ADA • BASIC ○ FACT • CORAL ○ LOT •
SIMSCRIPT • COMIT • ADAM •
FORTRAN ○ AESOP • COGENT
AIMACO • ALTRAN • JOVIAL ○ DPS ○
SNOBOL • META • DIAMAG • DYSAC
FLOW-MATIC • DYNAMO • FLAP ○
SMALLTALK ○ COBOL • LISP • PAL
BUGSYS • AMTRAN • GPSS ○ DAS
MAD • COURSEWRITER • ALGY • IT
FORTH • FORMAC • STRESS • ALGOL
PASCAL ○ C • PRINT • LOLITA ○
MAP • LOTIS • TMP ○ BASEBALL • GPL
PILOT ○ LOGO • PL/1 ○ DIMATE •

A collage of some of the names of computer languages that have been developed over the years.

Augustus De Morgan as her tutor. At the age of 19 she married Lord King (who later became Earl of Lovelace), and bore two sons and a daughter. Her husband was very supportive of her interest in mathematics, and she was fortunate to have the help of her husband, mother and servants in the upbringing of her children. This fact left her time to pursue mathematics. But how did she get involved in computer programming?

At age ten, Lovelace first met Charles Babbage. She went with a group of adults to view his laboratory and his wondrous machine, which had become a London attraction for society. She impressed Babbage because she was one of the few visitors to ask intelligent and thought provoking questions about his machine and his work. At age 21 she wrote Babbage asking for help in obtaining a tutor. One year later she undertook the task of translating one of the first papers written in French on the Babbage Analytical Engine. Her work did not end as a mere translation, but instead evolved into an extensive annotated translation which gave "a complete demonstration that the whole of the development and operations of analysis are capable of being executed by machinery" (in the words of Charles Babbage). Suffering from financial problems, especially when the government withdrew its support, the work of Babbage and Lovelace was sustained by her mathematical ability and her passion and enthusiasm for mathematics and the Analytical Engine. To keep the project alive she sold her jewels, and the two even used their mathematical talents in probability to raise money by trying to devise a system for winning at the horse races. But unfortunately Lovelace became ill with cancer, and died at the early age of 36 before completing her work.

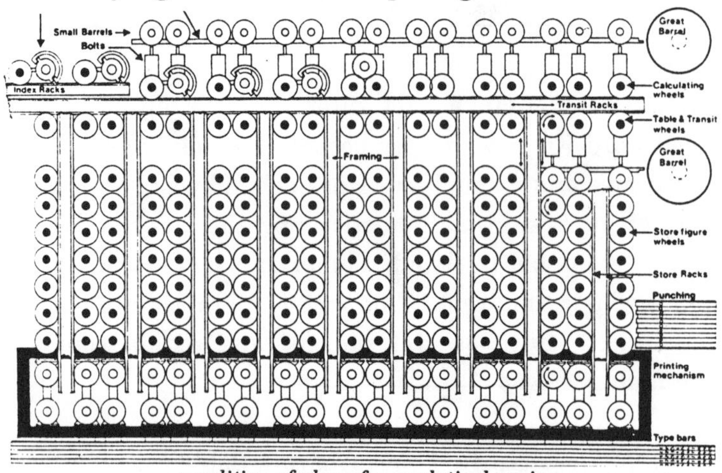

rendition of plans for analytical engine

Babbage's designs were definitely the predecessors of modern computers, and as testimony, IBM built a working model of the Analytical Engine as a tribute to Babbage and Lovelace. In honor of her programming work, the computer language ADA was named after her.

a work of Aristotle

This illustration is a section of a late manuscript of Aristotle's *Analytics*. This text points out the link between geometry and logic. *Universitjäts Bibiothek, Basel*

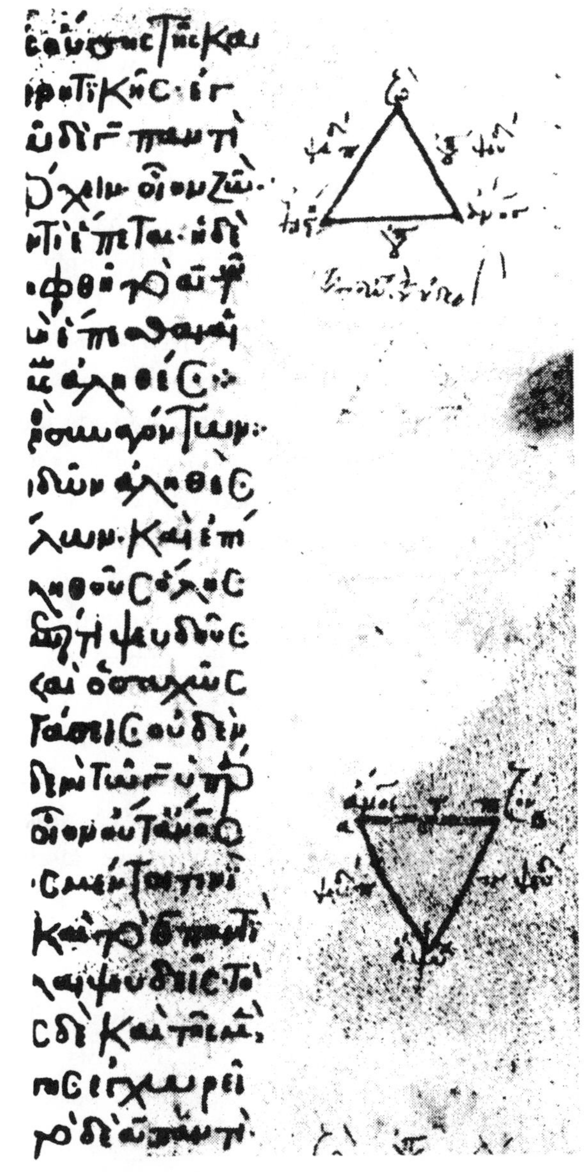

the camera obscura

The camera obscura has awed and captured the imagination of people for centuries. Even today with our high tech special effects, the camera obscura creates an image that is hard to believe. It has been used by astronomers, artists, photographers, inventors and magicians in all parts of the world.

Camera Obscura—meaning *dark room*— is an intriguing invention. One can speculate how the first camera obscura was invented. Someone entered a dark room which had a ray of light penetrating through some very small hole. Entering the room with a candle and closing the door, their candle was accidentally extinguished. The person then realized a beam of light was shining in the room, and projecting an image on the wall. To their astonishment it was a scene from outside the wall of the hole, except the image was upside down. Thus the first camera obscura must have been a darkened room or box, with a pin hole permitting the entrance of light. The light entering the pin hole projected the outside image upside down. Over the centuries various models were made. In the fourth century B.C. in China, the Mohists experimented with the camera obscura in their study of optics. But it was not until the eighth century A.D. that the camera obscura was used extensively in China as well as in Islamic countries. Leonardo da Vinci was fascinated by this device and in his drawings there is a diagram showing the operation of the camera obscura which he made in 1519. Portable versions were built, even pocket sized ones.

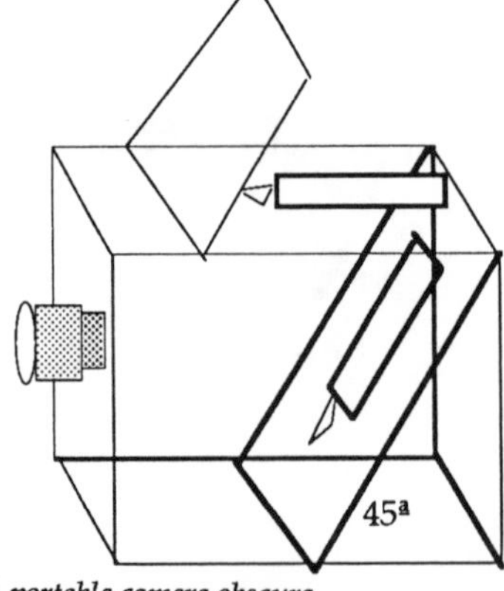

A portable camera obscura

Inside an early camera obscura. Today, a sophisticated version uses a parabolic screen and motorized optic unit.

By the 16th century artists adapted it to trace the projected image of their subject outside the camera obscura. By the 19th century, concave and convex lenses and mirrors were used to correct the inverted image. In 1826, Joseph Nicéphore Niépce using light sensitive paper projected an image on the paper with the camera obscura. The result—the invention of the photographic camera.

Sedan and tent models of camera obscura were used by artists.

Using scientific and mathematical ideas of optics and projection, the wondrous pinhole of light gadget was transformed into so many other uses. Astronomers used the camera obscura to to view eclipses of the Sun without harming their eyes.

Today one can experience the camera obscura at only a few locations world wide. The one located at the Cliff House in San Francisco, California has some special added features. Its "pin hole", optic unit, is located on its roof. Motorized, it rotates in 18° increments and reflects views of the outside through a mirror and a series of convex and concave lenses onto a parabolic screen, approximately four feet in diameter. Entering this darkened chamber and experiencing the camera obscura is almost magical. At first it seems you are viewing a motion picture, but immediately you realize it's a picture of what is actually happening outside. What

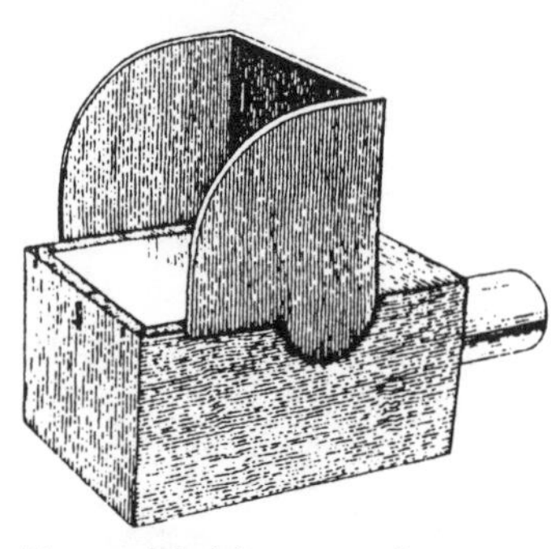

A box model of the camera obscura used around the 1830's.

Two Flagstaff, Arizona men with the four-person mobile camera obscura they constructed.

is truly amazing is that it is created by the projection of light entering a minute hole. When you witness it for yourself, you can appreciate the amazement of the discoverer of the first camera obscura; and you can understand why people may have believed it was magic or witchcraft of some sort.

a Greek computer?

In 1900, off Anti Kythera near Crete an ancient shipwreck was discovered. Along with pottery, marble works and bronze statues, an unusual bronze instrument was found in the wreckage. In 1951 Professor Derek de Solla Price of Yale University studied the device. He established

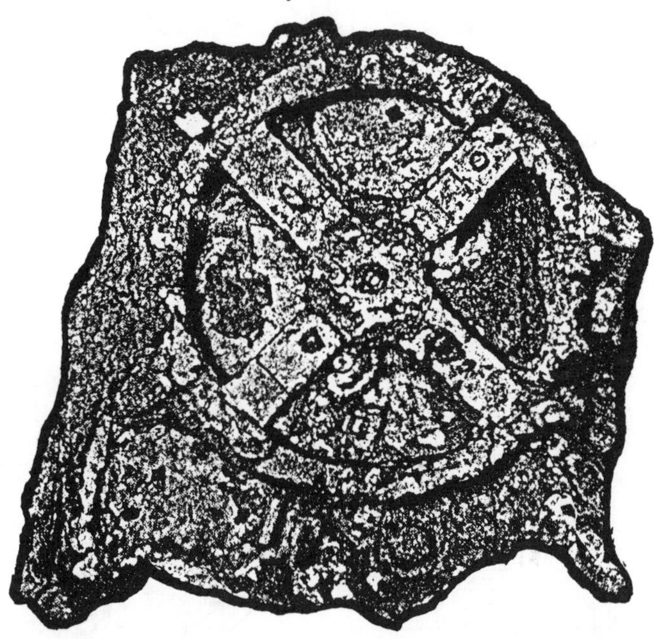

its time at about 78 B.C., and described it as *an ancient Greek computer,* able to indicate movements of the sun, moon and moon phases in the past, present and future. He depicted it as consisting of pointers, dials and more than thirty gears of various sized mesh on parallel levels with shafts rotating at different speeds. There is no mention in writings of this period of such a device, but a similar mechanism was mentioned by Cicero and later by others such as Ovid. Cicero describes a device made by Archimedes in the 3rd century B.C. which simulated the movements of the sun, moon and five planets.

mod arithmetic art

Some interesting designs are produced when the numbers from a modulo arithmetic table are transformed to a circle. The example here is produced from the numbers in the '2' row (this row is usually described as 19,2) of the MOD 19 table of multiplication.

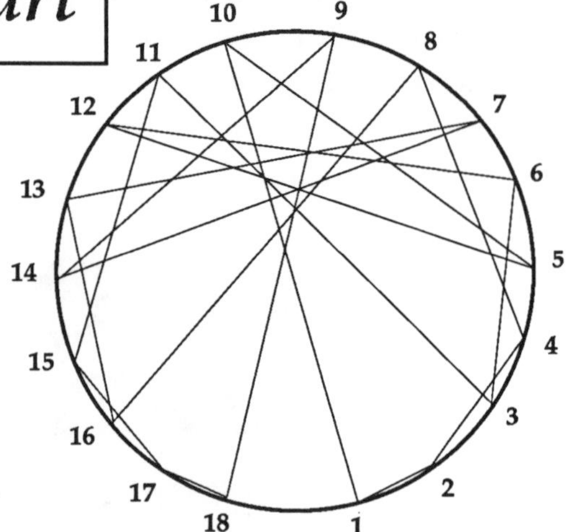

MOD 19 table of multiplication.

×	0	1	2	3	4	5	6	7	8	9	10	11	12	13	14	15	16	17	18
0	0	0	0	0	0	0	0	0	0	0	0	0	0	0	0	0	0	0	0
1	0	1	2	3	4	5	6	7	8	9	10	11	12	13	14	15	16	17	18
2	0	2	4	6	8	10	12	14	16	18	1	3	5	7	9	11	13	15	17
3	0	3	6	9	12	15	18	2	5	8	11	14	17	1	4	7	10	13	16
4	0	4	8	12	16	1	5	9	13	17	2	6	10	14	18	3	7	11	15
5	0	5	10	15	1	6	11	16	2	7	12	17	3	8	13	18	4	9	14
6	0	6	12	18	5	11	17	4	10	16	3	9	15	2	8	14	1	7	13
7	0	7	14	2	9	16	4	11	18	6	13	1	8	15	3	10	17	5	12
8	0	8	16	5	13	2	10	18	7	15	4	12	1	9	17	6	14	3	11
9	0	9	18	8	17	7	16	6	15	5	14	4	13	3	12	2	11	1	10
10	0	10	1	11	2	12	3	13	4	14	5	15	6	16	7	17	8	18	9
11	0	11	3	14	6	17	9	1	12	4	15	7	18	10	2	13	5	16	8
12	0	12	5	17	10	3	15	8	1	13	6	18	11	4	16	9	2	14	7
13	0	13	7	1	14	8	2	15	9	3	16	10	4	17	11	5	18	12	6
14	0	14	9	4	18	13	8	3	17	12	7	2	16	11	6	1	15	10	5
15	0	15	11	7	3	18	14	10	6	2	17	13	9	5	1	16	12	8	4
16	0	16	13	10	7	4	1	17	14	11	8	5	2	18	15	12	9	6	3
17	0	17	15	13	11	9	7	5	3	1	18	16	14	12	10	8	6	4	2
18	0	18	17	16	15	14	13	12	11	10	9	8	7	6	5	4	3	2	1

	1	2	3	4	5	6	7	8	9	10	11	12	13	14	15	16	17	18
2	2	4	6	8	10	12	14	16	18	1	3	5	7	9	11	13	15	17

These are elements from the '2' row of the MOD 19 multiplication table. To create the above design, the numbers from 1 to 18 are equally spaced around the circle, as in a clock. Then connect the corresponding numbers with line segments so—

1 with 2, 2 with 4, 3 with 6, 4 with 8, 5 with 10, 6 with 12, 7 with 14, 8 with 16, 9 with 18, 10 with 1, 11 with 3, and so on. Then shading in various ways produces artistic designs.

shapes & colors puzzle

Each shape comes in one of four shades. Place them in the grids so that every line of four has four different shapes and four different shades.

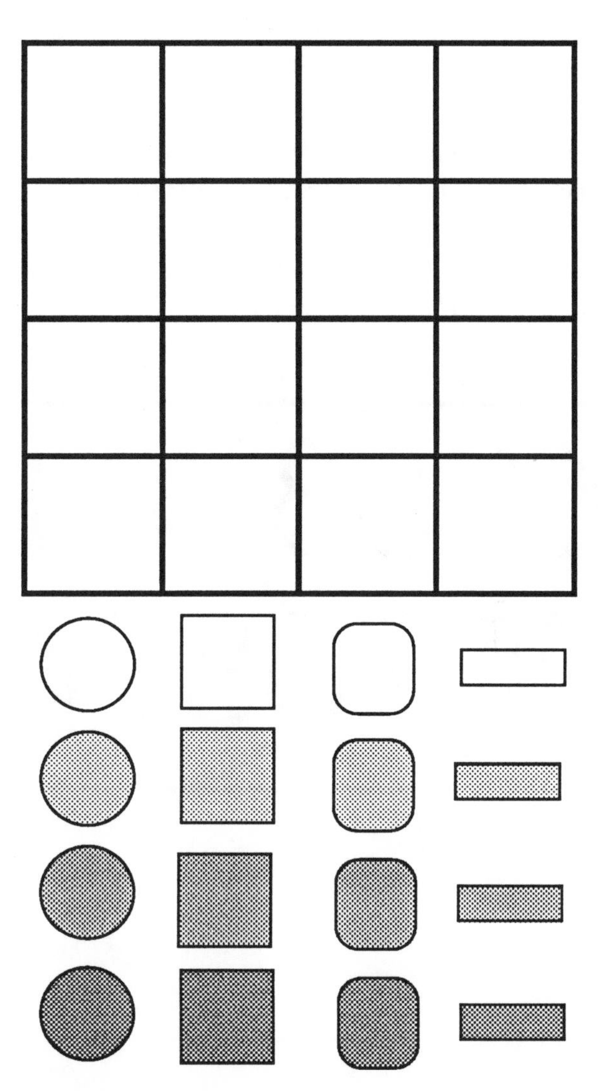

For solution, see the appendix.

$e^{\pi\sqrt{163}}$ = *integer?*

It seems amazing that these three irrational numbers — e, π and √163 — can be combined to form an integer. In fact, the Indian mathematician Srinivasa Ramanujan (1888-1920) first conjectured that $e^{\pi\sqrt{163}}$ was an integer. He felt he found its value to be— 262,537,412,640,768,743.999...

In 1972, computers had carried it out to 2 million places of 9's, but to be an integer one must know it repeats forever.

Finally, in 1974 John Brillo of the University of Arizona proved the number equals 262,537,412,640,768,744.

262537412
640768744

designs of the Pascal (arithmetic) triangle

A fascinating design comes to light when the numbers of the Pascal triangle are covered in the following fashion. If the number is odd, circle it with a pencil and shade it gray. When it is even, just circle the number. The pattern formed by the numbers in the arithmetic triangle enlarges itself and expands as one works down the triangle.

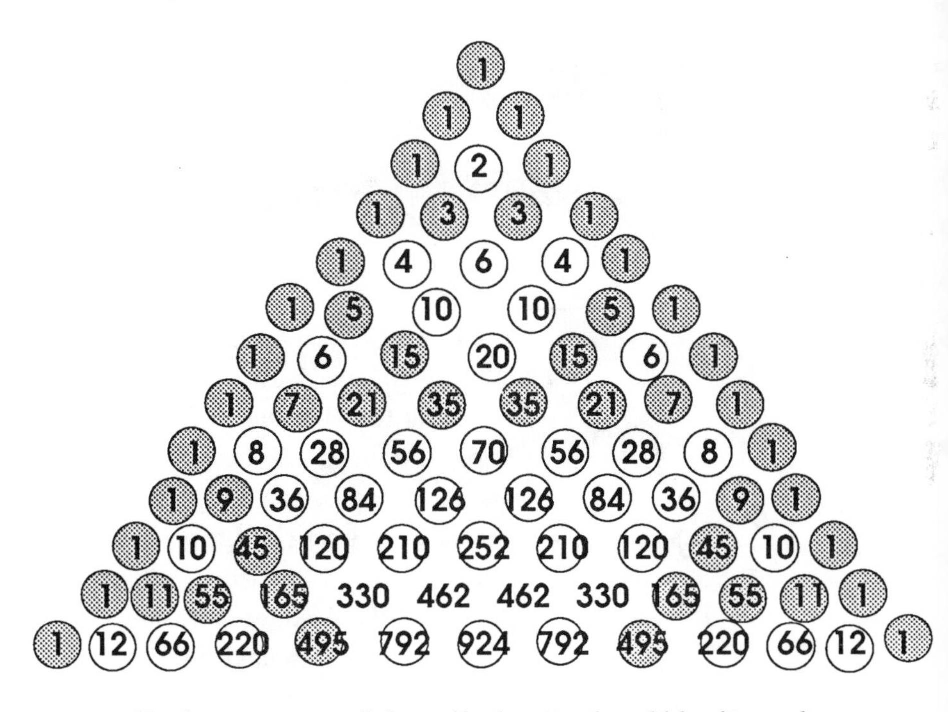

The above pattern can really be considered a pattern for multiples of two, and the pattern is more interesting when extended further. You may want to see what type of pattern one gets when other multiples are circled. Try multiples of 3.

the dock problem

There was a shortage of berths at this dock. These six boats had to be docked as shown. Planks run between them. Is it possible to take a walk from the dock and cross every gangplank only once and return to the dock? Test your networking on your answer.

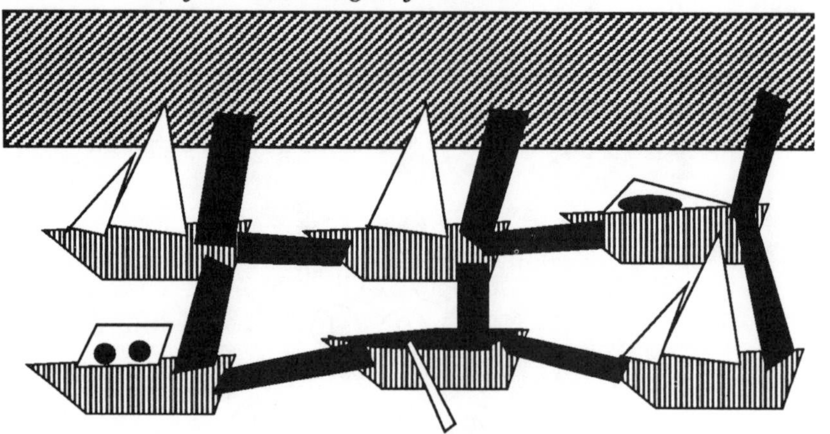

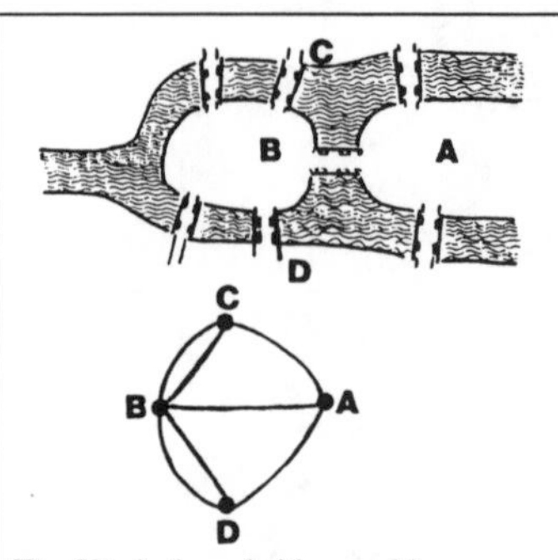

The Königsberg bridge problem & the network used to solve it

It was in 1736 when Leonard Euler (1707-1783) used the idea of networks to solve the Königsberg bridge problem. A network is a diagram used to solve a problem. For the Königsberg bridge problem Euler made a network which indicated the bridges with arcs, and points represented the land meeting the bridges. He reasoned that a traceable network could have at most two odd vertices — one for entrance and one for exit. As we see from its network, the Königsberg's bridge problem had 4 odd vertices, and was thus not traceable. The dock problem can be considered a type of a modern day Königsberg bridge problem.

For the solution, see the appendix.

Russian peasant multiplication

Though its origins are said to date back to ancient Egypt, today this method of multiplication is referred to as the Russian peasant method.

multiplying 158 by 39

158	~~39~~
79	78
39	156
19	312
9	624
4	~~1248~~
2	~~2496~~
1	4992

Total the numbers from column two that are not crossed out.

$$\begin{array}{r} 78 \\ 156 \\ 312 \\ 624 \\ +\ 4992 \\ \hline 6162 \end{array}$$

the answer

Procedure:

1) Place the two numbers to be multiplied at the heads of two columns.

2) The number in the first column is continually divided by two, ignoring remainders, until the number 1 is reached. The number in the second column is continually doubled, until it reaches the last row of the first column.

3) Now cross out all numbers in the second column that are in line with even numbers from the first column.

4) The product is given by adding the remaining numbers in the second column.

water jug problem

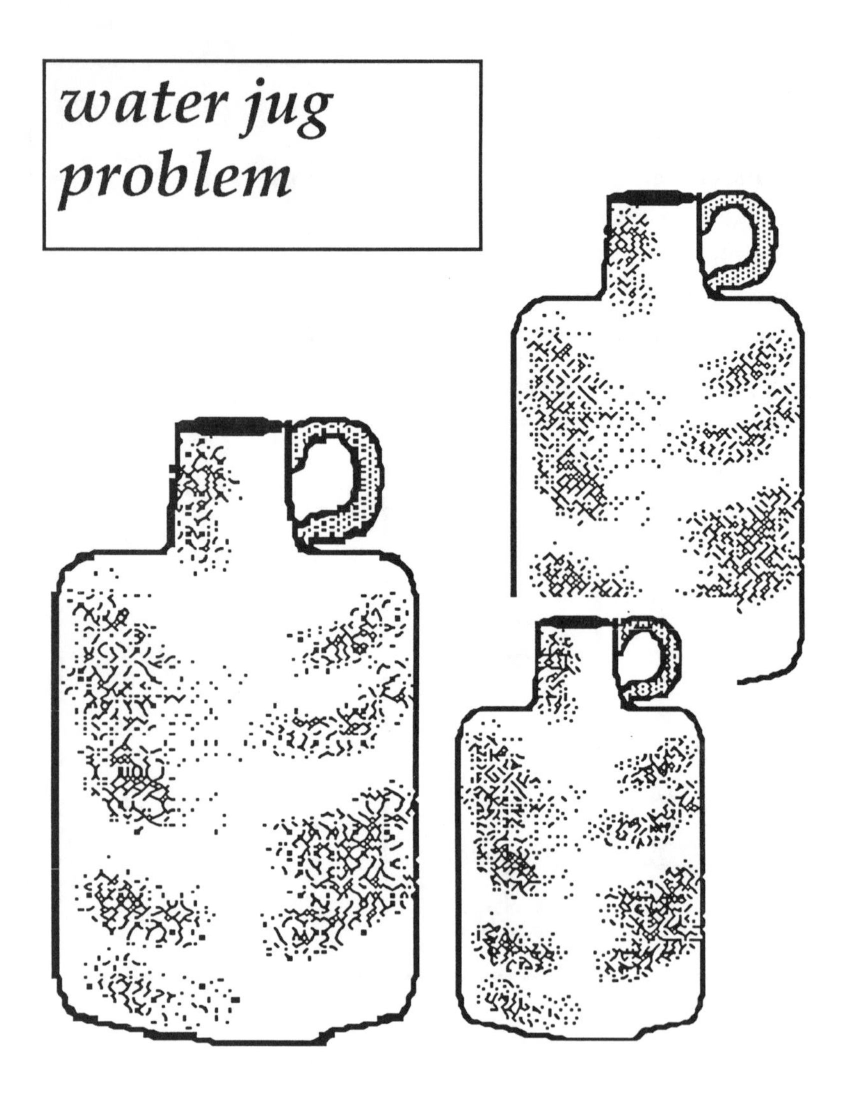

There is an 8 liter jug full of cider and two empty jugs—one is 3 liters and the other is 5 liters. The water is to be shared equally between two parties. How do they manage to each get 4 liters?

For solution, see the appendix.

Fibonacci magic

There are a number of "magic" tricks related to Fibonacci-like sequences. The Fibonacci sequence is 1,1,2,3,5,8,13,.... In it each term is the sum of the two preceding terms. Any sequence formed in this manner can be referred to as a Fibonacci-like sequence.

Pick any two numbers. Suppose you chose 5 and 7. Write as many new numbers as you like by adding the two preceding numbers

5
7
12
19
31
50
81
131
212

343
555
898

Draw a line between any two of the numbers listed. The sum of the numbers above the line will always be the second ***number below the line minus the second number from the beginning.*** *In this case the sum would be 555-7=548.*

Discover why this always works.

Kepler's derivation for the area of a circle

Johannes Kepler (1571-1630) developed a very interesting method for explaining how the area formula of a circle was derived.

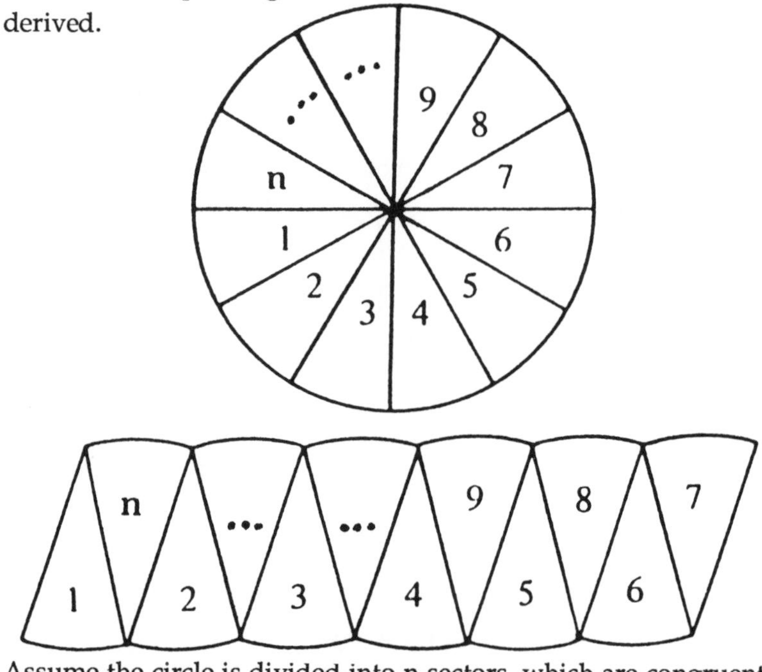

Assume the circle is divided into n sectors, which are congruent isosceles triangles. Since these isosceles triangles originally come from a circle, each has an altitude equal to the radius of the circle. When placed together, as in the diagram, they form a type of parallelogram shape. The area of a parallelogram is found by multiplying its base times its altitude. In this case the base is the length of half the circle, namely (1/2) the diameter times π or $(1/2)d\pi=(1/2)2r\pi=r\pi$. The altitude of this parallelogram is the same as the altitude of one of its composing isosceles triangles, namely r. Thus, the circle's area=the parallelogram's area$=(r\pi)r=\pi r^2$.

the game of doublets

Lewis Carroll (1832-1898), the author of *Alice in Wonderland,* was better known to his colleagues as mathematician Charles L. Dodgson. Besides *Alice in Wonderland* and *Alice through the Looking Glass,* he also wrote a number of mathematics books. The influence of mathematics can be found in most of Carroll's work. He definitely had a knack for presenting mathematical ideas in a recreational form.

Doublets is a word game Carroll created which became very popular. It begins with two words the same length, for example CAT and DOG. Change a word by forming other successive words of the same length having altered only one letter at a time.

CAT
BAT
BAG
BOG
DOG

The object of the game is to make the transformation in the least number of words.

The English magazine *Vanity Fair* ran a contest of the *Game of Doublets.* Their first competition was—

PROVE GRASS TO bE GREEN
EVOLVE MAN FROM APE
RAISE ONE TO TWO
CHANGE BLUE TO PINK
MAKE WINTER TO SUMMER
PUT ROUGE ON CHEEK.

You may want to give it a whirl. Have fun!

musical scales— *mathematics to the ears*

The speed of light, c, π, e, φ and Avogadro's number are all examples of constants of our universe. These are numbers which have vital roles in equations and formulas which define various objects of our world — be they geometric, physical, chemical or commercial. Among these famous constants, the concept of an octave should be included as a constant of a special nature. The octave plays a very important part in the world of music. It establishes the unit or distance of a scale. Just as the ratio of a circle's circumference to its diameter always produces the constant π, the ratio of the number of vibrations of a plucked string to a string half its length is the ratio 1/2. These notes have the same sound, and form the length of an octave.[1] The shortened string vibrates twice the amount per second than the original string. The number of notes or subdivisions of an octave is arbitrary. It is influenced by a number of factors. Consider the various factors that come into play when creating a particular scale. Sounds or notes are what comprise a scale. Each has a particular frequency.[2] As mentioned, two notes are an octave apart if the frequency of one is double that of the other.[3] The trained ear can hear about 300 different sounds or notes in one octave. But to produce a scale with this many notes would be ludicrous, since traditional instruments cannot produce that many notes. For example, if there were 300 notes in an octave, a piano of eight octaves would have 2400 keys. Can you imagine a pianist running back and forth along the keyboard? Thus, the number of possible notes is restricted by the physiology of our ears and by the capabilities of our instruments. How and which of the 300 discernible sounds were selected for a scale? Selecting the notes of a scale is analogous to selecting a numeration system. —What base should be used and which symbols should be made to represent the numbers? — With the scale, the length of an octave string needed to be selected and the number of subdivisions (notes comprising the scale) had to be determined. As with numeration systems we find these evolved differently in various civilizations. The ancient

Greeks used letters of their alphabet to represent the seven notes of their scale. These notes were grouped into tetrachords (four notes), which were put into groups called modes. The modes were the forerunners of modern Western major and minor scales. The Chinese used a pentatonic (five note) scale. In India, music was and is improvised within specific boundaries defined by ragas. The octave is divided into 66 intervals called srutis, although in practice there are only 22 srutis, from which two basic seven-note scales are formed. The Persian scale divided the octave into either 17 or 22 notes. We see that although the octave was a given constant, different musical systems evolved. In addition, the musical instruments from one culture cannot necessarily be used to perform the music of another.

* * *

Archeological artifacts of instruments, vases, statues, and frescoes depicting vocal and instrumental musicians have been unearthed. There are many early examples of written music — Sumerian clay tablets excavated in Iraq appear to show an eight note scale (circa 1800 B.C.); fragments written on stone carvings and pieces of papyrus from ancient Greece, Greek text books (circa 100 A.D.); a Greek manuscript in which notes were written using their alphabet (circa 300 A.D.); manuscripts of a Moslem-Arabic chant from 8th century Spain.

During the 6th century B.C., Pythagoras and the Pythagoreans were the first to associate music and mathematics. The Pythagoreans believed that numbers, in some way, governed all things. Imagine their delight when they discovered the octave of a note, the periodicities of notes, and the ratio of notes on string instruments. They discovered the connection between musical harmony and whole numbers by recognizing that the note caused by a plucked string depended upon the length of the string. They found that an entire scale could be produced by taking integral ratios of a string's length.[4]

Are musical scales needed to produce music? If they were, how would birds sing? Yet most verbal renditions of a story or a musical tune change slightly with each oral communication. To have a composition performed, scales are essential. Scales are the written language of music, even as equations and symbols are the written language of mathematics.

[1]The term octave comes from the Latin word for 8. A diatonic scale, has seven distinct notes from C to B and high C, the octave, is the eighth note.

[2]Frequency is the number of vibrations per second. There is a one-to-one correspondence between sounds and numbers (its frequencies), although our ears are not designed to discern all the possible distinct sounds.

[3]A plucked string produces a certain note, e.g. C has 264 vibrations per second, when the string is depressed at half its length its octave is produced, which vibrates at 528 vibrations per second.

[4]For example, by starting with a string that produced the note C, then 16/15 of C's length produced B, 6/5 of B's produced A, 4/3 of A's produced G, 3/2 of G's produced F, 8/5 of F's produced E, 16/9 of E's produced D and 2/1 of D's gave low C. In addition, they believed that planets had their own music, that the celestial bodies produced musical sounds. This idea came to be known as the "music of the spheres". Kepler believed in the music of the spheres, and in fact wrote music for each of the the known planets. Today, astronomers have received radio signals carried by solar winds. These sounds, including whistles, pops, whines, hisses which when synthesized at increased speeds, are more melodious. Scientists have also observed oscillations from the sun which they surmise produce vibrations that are in various periods.

dynamic rectangles

Dynamic rectangles are rectangles that can be generated from the unit square, such as those illustrated below. Included in this family of rectangles are the golden rectangles and the root rectangles. What is especially exciting about these rectangles is that each of them forms its dynamic spiral (a logarithmic spiral) which has many applications in nature (especially in the area of growth patterns) and in art designs (in designs requiring special proportions and ratios).

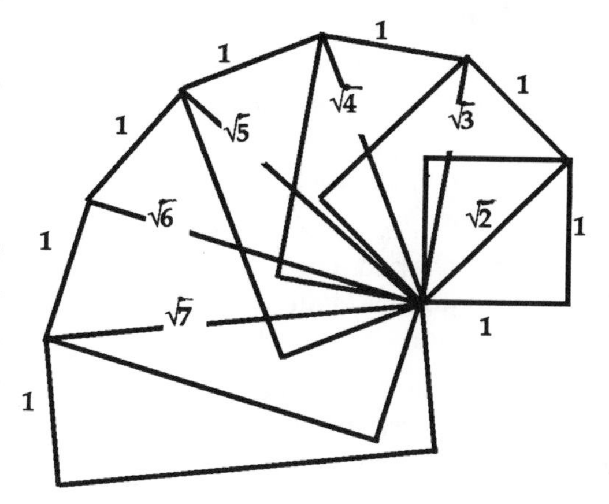

Each dynamic rectangle also produces a rectangular spiral. Its formation is illustrated below in the $\sqrt{2}$ rectangle.

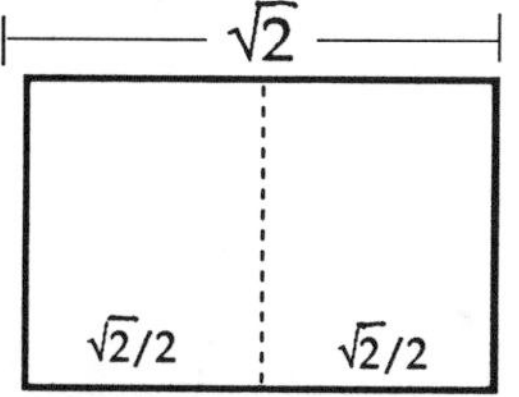

A rectangular spiral for $\sqrt{2}$ is illustrated in the diagram below in the right hand corner.

In the rectangular spiral with sides 1 by $1/\phi$ (i.e. the reciprocal of the golden mean $\approx 1/1.618\ldots \approx .618034$) the length of the spiral is equal to the golden ratio, ϕ.

The $\sqrt{2}$ dynamic rectangle is especially of interest because it is the only one in which its half is similar to the whole rectangle.

designs made from dynamic rectangles

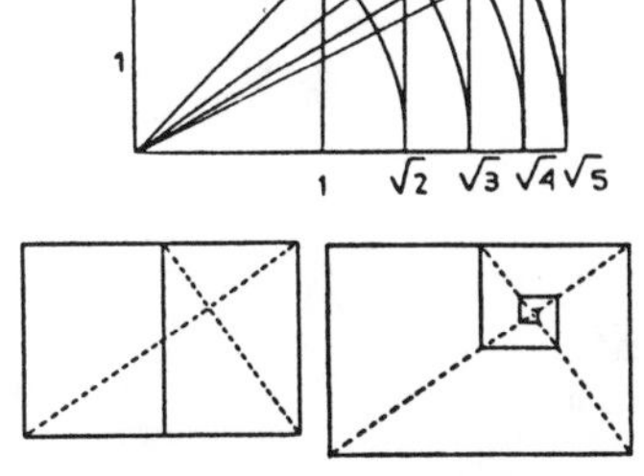

a rectangular spiral

creating irregular mathematical mosaics.

In recent years mathematicians have discovered and designed various way to tessellate a plane with different polygons. Here are some of their findings:

•*Convex polygons with more than six sides cannot tessellate a plane.*

•*For 3-sided convex polygons*
Any shaped triangle can be used to tessellate a plane.

•*For 4-sided convex polygons*
Any shaped convex quadrilateral can be used to tessellate a plane using rotation, reflection and translation.

•*For 5-sided & 6-sided convex polygons*
Only certain convex pentagons and hexagons tessellate. The convex hexagons and pentagons have be analyzed and classified. Some examples are:

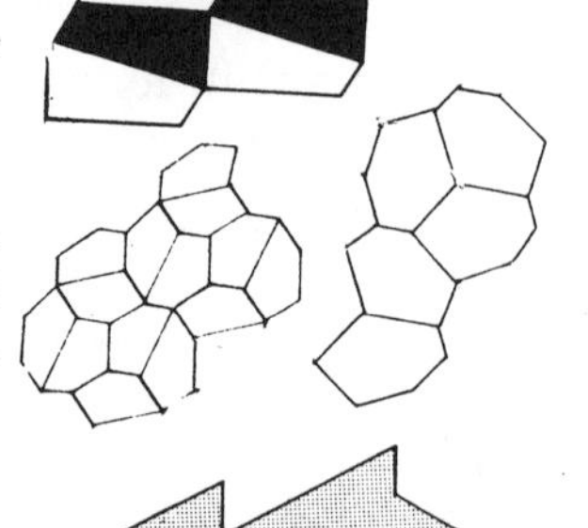

•*For nonconvex polygons*
To make matters even more interesting consider nonconvex polygons. Much study is under way with congruent nonconvex polygons, such as pentaminoes or any polyominoes[1]. Or what about polyiamonds[2] or polyhexes[3]. Much is still left to be resolved. One thing is certain, many beautiful mathematical mosaics will to be created!

---

[1] Polyominoes are formed from congruent squares. For example a pentomino is formed by joining together five congruent squares in various ways.

an example of a pentamino tessellation

[2] Polyiamonds are formed from congruent equilateral triangles.

examples polyiamonds

[3] Polyhexes are formed from regular hexagons.

examples of trihexes that can independently tile a plane

encircling the Earth

Suppose a piece of rope exactly encircles the Earth at the equator. If an additional yard is added to the rope, will it be noticeable? How far above the earth will this new rope stand along each point?

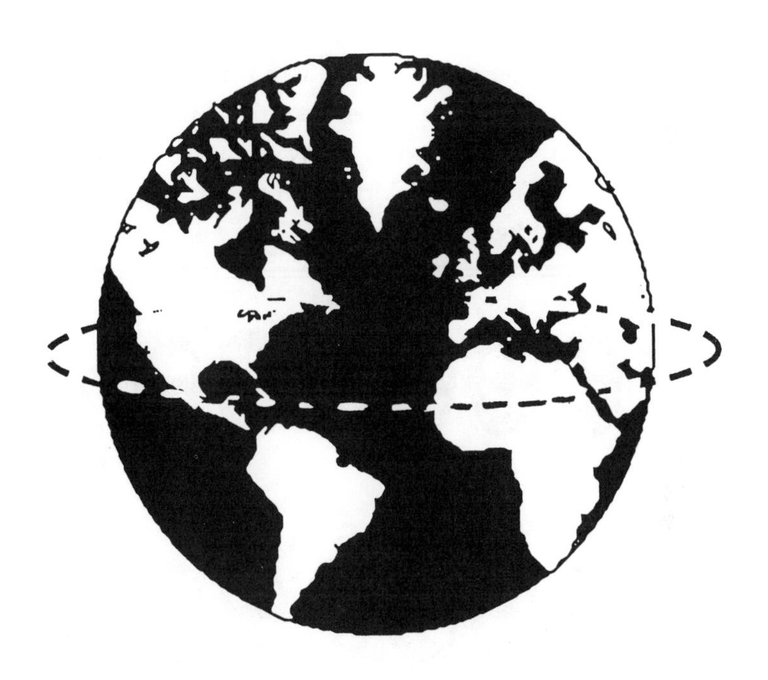

*Assume the circumference of the equator is 25,000 miles long.

For solution, see the appendix.

the game of mancala

Mancala is the generic name for a family of games that originated in Egypt over 3500 years ago. In the excavations of the pyramid of Cheops and other temples of Egypt, game boards were found carved out of stone. For over a thousand years, the Moslems played a major role in the global spread of mancala, introducing it to the regions they conquered. It derives its name from the Arabic word, *naqala* which means to move. Later, African slaves brought the game to Surinam and the West Indies. The game was widely played in the cafés of Cairo in the 19th century. It was there that Europeans learned mancala. Today it is played in many countries of the Middle East, South East Asia and Africa. It is probably the most continuously played game in history.

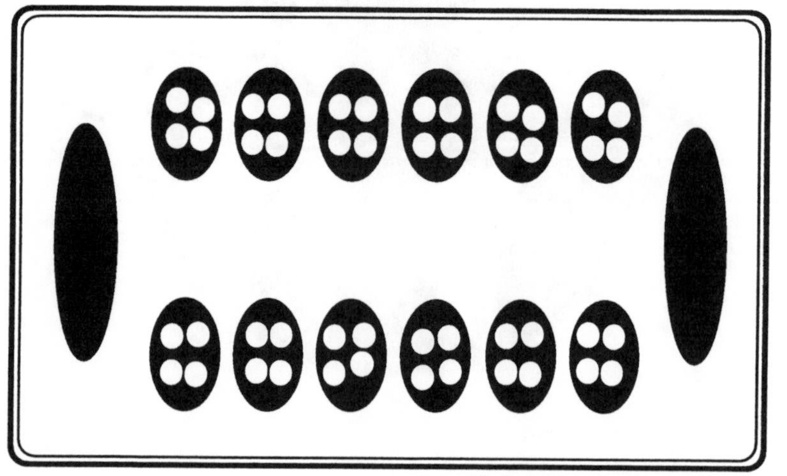

The beginning set up for mancala.

Its existence for so many years accounts for the over 200 versions of the game. One of the beautiful aspects of mancala is that it can be played anywhere, with or without a formal wooden or ceramic board and pieces. Its board can be improvised easily in sand, earth, or on paper. Stones, beads, beans, shells or coins can be used as pieces. Mancala is played by two people.

Here is a version of mancala that is played in East Africa.

the board

The board consists of twelve scooped out wells for the playing area and two other wells for storing captured pieces.

Four pieces are placed in each of the twelve wells.

the moves

From his or her side, a player chooses a well's pieces to move. Moving clockwise, the player places a piece in each well after the emptied well, until the pieces in his or her hands are exhausted. No well can be skipped along the way, nor can a well receive more than one piece. If a well has, for example 14 pieces, the player places a piece in each well while proceeding counter-clockwise around, even onto the opponent's side, but the player *must pass over* the emptied well from which the pieces were taken.

object of game

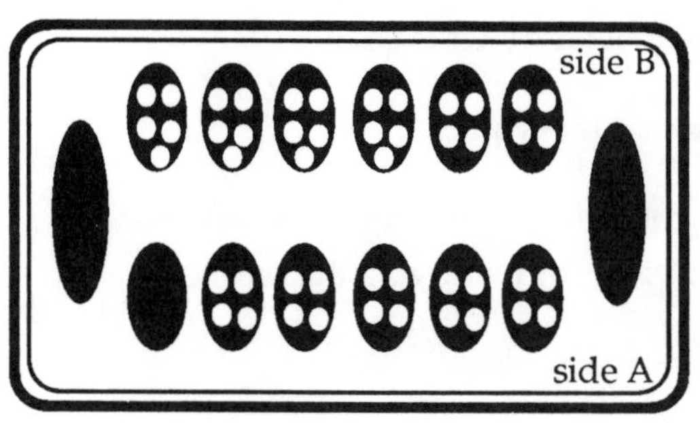

the above board shows a first move by player A

The object of the game is to capture pieces from your opponent. Pieces are captured by having the last piece from your hand go into a well that contains either 1 or 2 pieces on your opponent's side. You capture these pieces and place them in your reservoir. After capturing the pieces in an opponent's well, the player is also entitled to take the pieces from the previous well if it is on the opponent's side and has 2 or 3 pieces in it. You can continue to capture pieces in your opponent's consecutive wells as long as the wells contain 2 or 3 stones.

special rules

A player is not allowed to empty all the opponent's wells, even though the play allows it. A player must always try to leave the opponent a move.

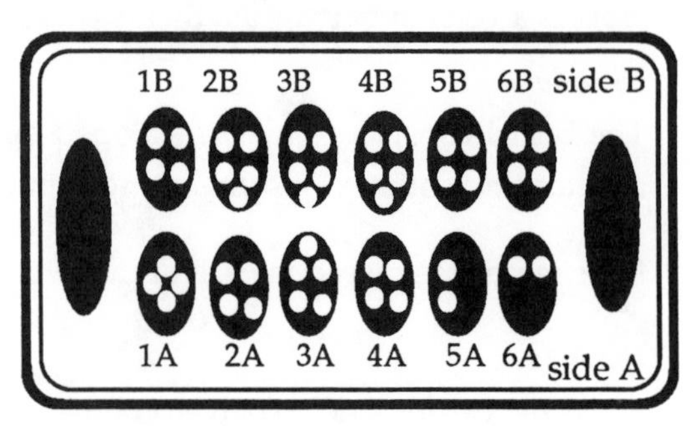

For the board above, player B can move his or her pieces from well 3B and capture pieces in well 5A and also pieces in the previous well 6A.

If your opponent's wells are already empty , you must try to leave at least one stone in a well of your opponent during your move.

end of game

The game ends when all the wells on a player's side are empty and it is his or her move.

The player with pieces left on his or her side then places them in the reservoir and includes them in his or her total.

The player with the most pieces is the winner.

As has been done over the centuries, mancala can be adapted and changed by adding more pieces or modifying the rules.

Egyptian fractions & the eye of Horus

Egyptians developed a method to write fractions that had a numerator of 1. The Egyptian symbol, ◎, for mouth was used to write these fractions. When this symbol was used with a numeral it meant "part", and it was placed above the numeral much as a line segment when we write the fraction 1/3. The Egyptian one-third would have been written—

$\frac{1}{3}$ $\frac{1}{5}$ $\frac{1}{10}$ $\frac{1}{100}$

In order to write a fraction whose denominator was not 1, for example 3/5, they would rewrite it as 1/2 + 1/10. For frequently used fractions, they had developed special glyphs—

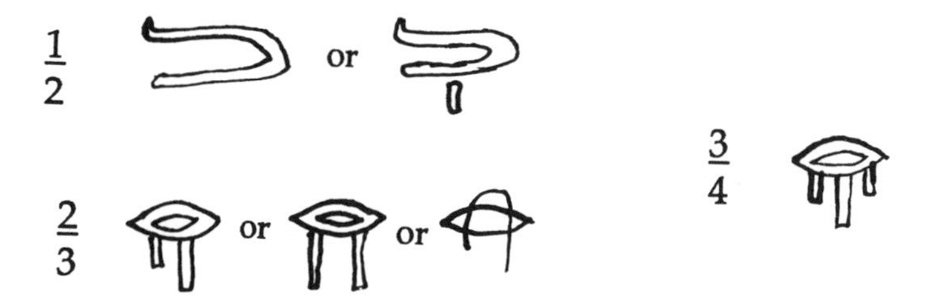

When fractions were needed to express volumes for liquids, grain and agricultural products, the parts of the eye of Horus were designated with specific fractional values. The origin of this practice is tied to the myths of the gods Nut, Isis, Nephthys, Thoth, Set, Osiris and Horus. The diagram illustrates which fractional values— 1/2, 1/4, 1/8, 1/16, 1/32, 1/64— the Egyptians associated with each part of the eye of Horus.

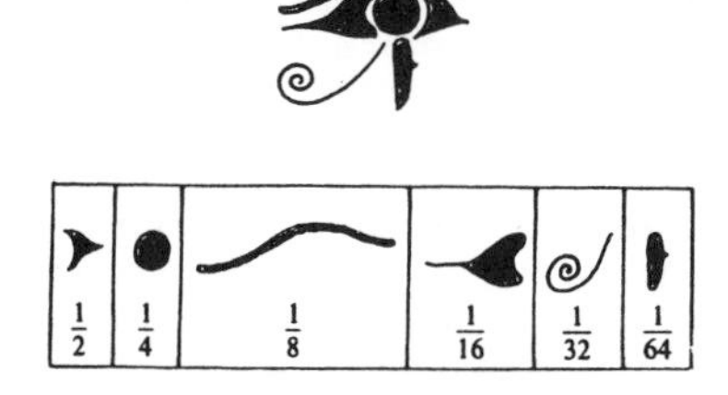

Pascal's amazing theorem

The famous French mathematician Blaise Pascal* (1623-1662) discovered and proved the theorem—

A six sided polygon is inscribed within ***any*** *conic section. When its sides are extended into lines— AB, BC, CD, DE, EF, AF—three points of intersection are created, namely, P, Q, and R. These three points are* ***always collinear.***

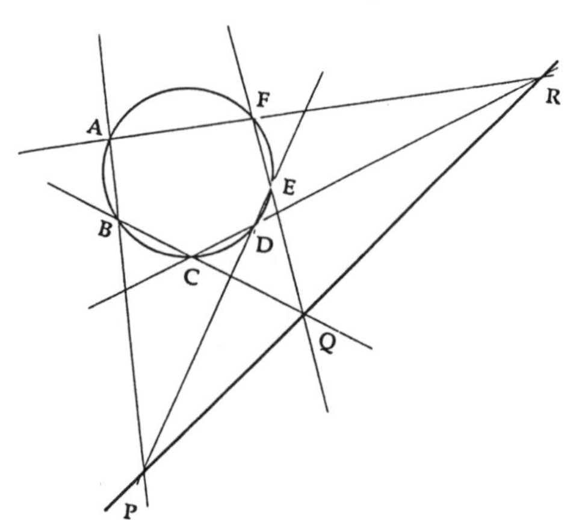

Using this theorem, Pascal was able to deduce much of what we know about conic sections! Nearly 400 corollaries to this theorem were proven.

In addition, using the idea of projection (developed by Gérand Desargues), Pascal illustrated the theorem for other conics besides the circle.

---

*Pascal is credited with making many theoretical mathematical and scientific discoveries. His works include extensive discoveries with the Pascal triangle, with hexagrams, with the theory of liquids and hydraulic pressures, and with the theory of probability. In addition, he invented an adding machine at the age of 18. But in 1654, his conversion to Christianity essentially ended his work in mathematics. These excerpts from his essay, ***Thought,*** reveal his use of many mathematical references and examples for his various arguments and points he is making. "*...Expand our conception as we may beyond imaginable space, we beget only atoms in comparison with the reality of things. It is an infinite sphere whose center is everywhere, its circumference nowhere. In brief, it is the greatest sensible mark of the omnipotence of God, that our imagination loses itself in this thought. ...All things proceed from nothing and are borne on to the infinite. Who can follow these amazing processes? The author of these wonders comprehends them; no one else can.....I know people who cannot comprehend that if one takes four from zero, zero is left. Too much pleasure disturbs; too many concords in music displease; too many benefits irritate; we wish to have the means of overpaying the debt.*"

puzzles to exercise the mind

Dissection puzzles: Problems involving the dissection of a rectangle into a square date back to the 18th century and French mathematician Jeanne Étienne Montucla. These particular types of problems were expanded by the works of American puzzlist Sam Loyd (1844-1911) and British puzzlist Henry Dudeney (1847-1930).

Cut this rectangle in two pieces, and form a square from the pieces.

4

9

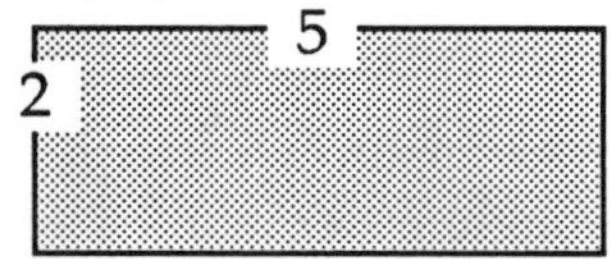

Cut this rectangle in three pieces, and form a square from the pieces.

Henry Lingren is one of today's specialists in these kinds of problems. He has written an entire book of them, titled *Geometric Dissections*.

stick puzzle:
Remove five match sticks and end up with only three squares that are the same size as the original ones.

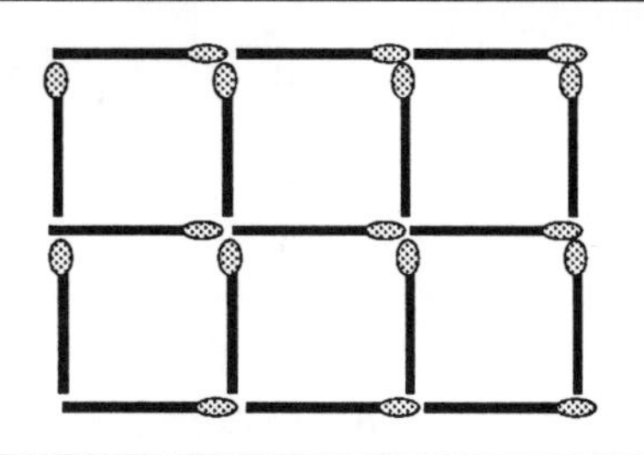

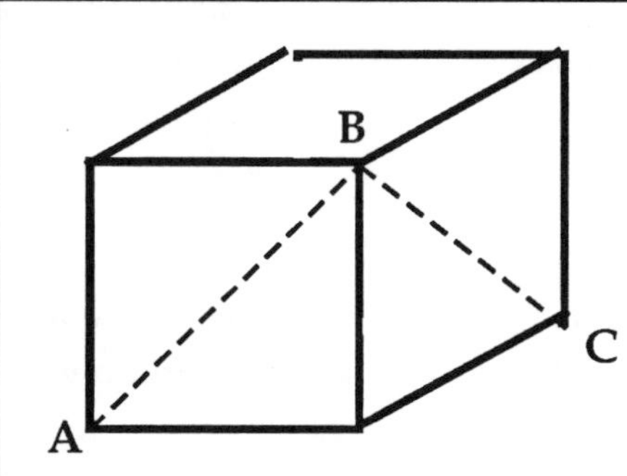

cube puzzle:
The solid shown is a cube. Determine the measure of angle ABC.

For solutions, see the appendix.

mathematics & origami

A square is transformed into a box.

A square is transformed into a bird.

A square is transformed into a snake.

A square is transformed into an elephant.

Unless one knew better, one would think we were talking about a magic show or topology[1].

An origami-like sculpture at the United Airlines terminal, San Francisco Airport, California.

Origami is an art form that dates back to 583 A.D. when Buddhist monks brought paper into Japan from China through Korea. Since the manufacturing of paper at that time was costly, people used it with care, and origami became an integral part of certain ceremonies. The art of origami has been shared and passed on from generation to generation. Animals, flowers, boats, and people have all been created with origami. (The word origami is derived from *ori—to fold* and *gami—paper.*) Origami has delighted and frustrated enthusiasts over the centuries. In fact, today there are international origami societies established in Britain, Belgium, France, Italy, Japan, the Netherlands, New Zealand, Peru, Spain, and the United States.[2]

In creating an origami figure, the origamist begins with a square sheet of paper and transforms it into any shape limited only by his or her imagination, skill and determination.

A square was probably chosen as the original starting unit of origami because, unlike the rectangle or other quadrilaterals, it possesses four lines of symmetry. Although some other regular polygons and circles have more lines of symmetry, they lack the right angles of the square, and would have been more difficult to manufacture. Sometimes origamists do begin with other units, but the purists work with squares without using glue or scissors.

A study of the creases impressed on the square sheet of paper, after an origami object has been created, reveals a wealth of geometric objects and properties.

This diagram shows the creases that were impressed on a square when it was folded into a flying bird.

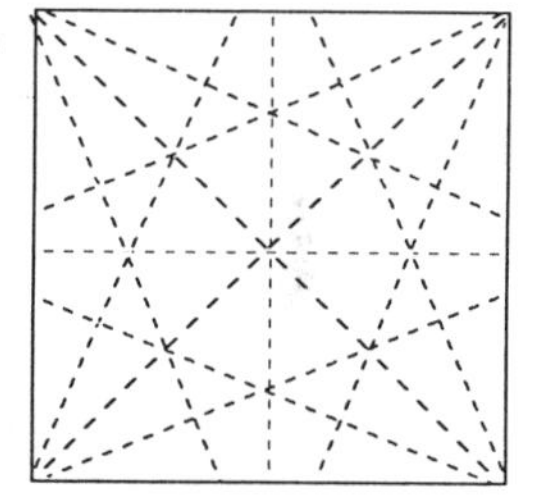

The creases on a square can illustrate the mathematical ideas of similarities, lines and points of symmetry, congruences, ratio and proportions of shapes, and iterations (continual repetitions of patterns within patterns) resembling the formation of geometric fractals.

Studying the progression of an origami creation is very enlightening. One begins with a square (a 2-dimensional object), and then manipulates the square to form a figure (a 3-dimensional object). If it is a new creation, the origamist will unfold the figure and study the creases impressed on the square. This process involves moving between dimensions. The creases represent the object's 2-dimensional projection onto a flat plane, namely the square. A transformation of a 2-D object to a 3-D object and back is related to the field of projective geometry.

In the book *Folding the Universe,* author Peter Engel, a master of the art and science of origami, reveals his years of work, unique discoveries and creations in origami. Engel has taken origami to a whole new plateau. He emphasizes the strong connection between origami, mathematics, and nature by drawing analogies to minimization problems, fractals and the chaos theory. An origami creation begins with a finite amount of material (e.g. a square of fixed dimensions) and evolves into a desired form, not unlike the restrictions placed upon nature in the formation of natural forms, such as bubbles.

Origami is experiencing a renaissance. It has come a long way from the foundations developed by early paperfolders. The complexity of the figures folded by today's masters are truly amazing. Their skill in transforming a square sheet of paper, without the use of scissors or glue, is incredible. The completed forms are not simple boxes or flowers, but anatomically accurate animals, realistic lifelike paper sculptures — squid, spiders, snakes, dancers, furniture. To achieve such proficiency and creativity takes years of work, experience and study — it is analogous to the years which such artists as M.C. Escher devoted to developing the art of tessellation.

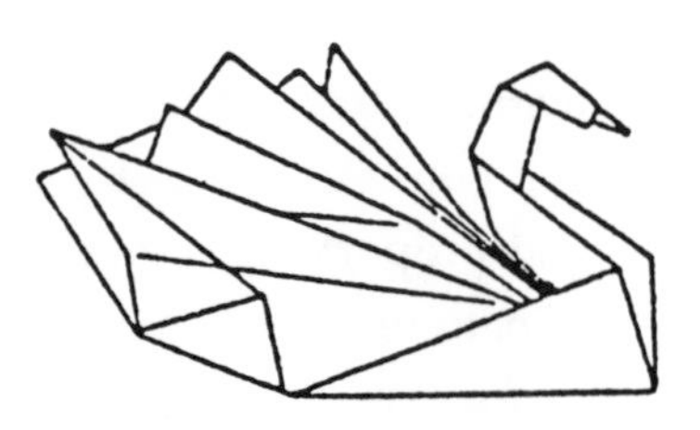

The mathematics, whether identified as such by origamists, is there. Like tessellation art, the understanding of mathematics enhances one's ability and creativity.

---

[1]Topology is a special kind of geometry that studies properties of an object that remain unchanged when the object is distorted by being stretched or shrunk. Unlike Euclidean geometry, topology does not deal with size, shape or rigid figures. This is why topology is often referred to as rubber sheet geometry. Imagine objects existing on a rubber sheet that can be stretched and shrunk. In the process of these transformations, one studies the characteristics that remain unchanged.

[2] The Friends of Origami Center of America is located at 15 West 77th St., New York, NY 10024,

The British Origami Society is located at 12 Thorn Road, Bramhall, Stockport, Cheshire SK7 I HQ.

Sam Loyd's missing number puzzle

The eight legible digits remaining in this archaeologist's discovery furnish enough information to determine the missing digits. What are the missing digits?

For solution, see appendix.

paradoxes

"... Major paradoxes provide food for logical thought for decades and sometimes centuries."
—Nicholas Bourbabki

"Do not read anything on this page." Paradoxes such as this one are statements which either appear to contradict themselves, create senseless or astonishing conclusions, or form an endless circle of reasoning. Over the centuries paradoxes have fascinated and perplexed the logical mind. Intriguing and at the same time disturbing, they appear in a broad spectrum of disciplines, including literature, science and mathematics, as well as in our everyday encounters, such as the photograph on this page. Regardless of the type of paradox, the confusion and questions created can be both interesting and enjoyable. In particular, mathematical paradoxes become realms of discoveries. A brief exploration of some well known paradoxes will amuse even as they tax and exercise our minds.

A homeowner, irritated with all the garage sale notices, posted this paradox.

* * *

In the 5th century B.C., Zeno, using his knowledge of infinity, sequences and partial sums created these well known paradoxes:*

The Dichotomoy — A traveler is walking to a specific destination. He or she must first walk half the distance. Then the traveler must walk half the remaining distance. Then half of the part that remains. There will always be half of the part that remains to walk, therefore the traveler will never reach the destination.

Achilles and the tortoise— In a race between Achilles and the tortoise, the tortoise was given a 1000 meters head start, because Achilles could run 10 times faster than the tortoise. When the race started and Achilles had gone 1000 meters, the tortoise was still 100 meters ahead. When Achilles had gone the next 100 meters, the tortoise was still 10 meters ahead. Zeno argued that Achilles would continually gain on the tortoise, but he would never reach him. Was his reasoning correct?

Eublides' paradox—The Greek philosopher Eublides (4th century B.C.) argued that one could never have a pile of sand. He proposed that certainly one grain of sand does not constitute a pile of sand. And if one adds another grain of sand to the one, they do not make a pile. He said that if one does not have a pile of sand and if by adding one grain to what one has, one still does not have a pile; then one will never have a pile of sand. He is also credited with the paradox— "The statement I am making is false."

Epimenides' paradox—Epimenides was from Crete and his paradox simply stated "All Cretans are liars. "

Aristotle's wheel paradox —Two concentric circles are shown on this wheel. The wheel moves from A to B as it rotates once. Notice |AB| corresponds to the circumference of the large circle. Since the small circle also rotates once and travels the distance |AB|, is not its circumference |AB|?

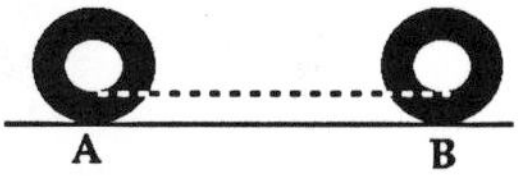

A B

A coin paradox—The top penny is moved halfway around the penny below it. It ends up in the same position in which it originally started. Since it traveled half of its circumference, one would expect it to end upside down. Can you explain what happened?

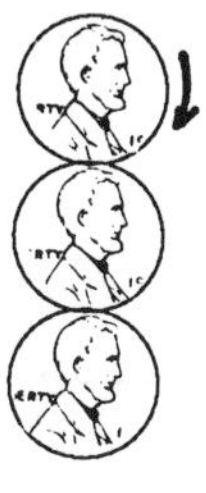

algebraic paradoxes—Algebraic paradoxes are numerous. Here are a few well known ones to ponder.

If a=b, then 1= 2.

proof

$a=b \rightarrow a^2=ab \rightarrow a^2-b^2=ab-b^2 \rightarrow$

$(a^2-b^2)/(a-b)=(ab-b^2)/(a-b) \rightarrow$

$(a-b)(a+b)/(a-b)=b(a-b)/(a-b) \rightarrow a+b=b \rightarrow$

$a+a=a \rightarrow 2a=a \rightarrow \therefore 2=1$ or $1=2$

proof of 1=-1

$-1 = (\sqrt{-1})^2 = (\sqrt{-1}) \bullet (\sqrt{-1})$

$= \sqrt{(-1)(-1)} = \sqrt{1} = 1$

barber paradox —This paradox dates back to 1918. In a particular village, the barber shaves all those in the village who do not shave themselves. Who shaves the barber?

Walt Kelley paradox— "We have met the enemy, and he is us."

Oscar Wilde paradox— " The only way to get rid of temptation is to yield to it."

Paradox from ***Don Quixote*** —Sancho Panza is governor of the island of Barataria. Persons coming to the island must state why they are coming. If they tell the truth, they will go free. If they lie, they will be hanged. One day a traveler arrives and states "I am here to be hanged." What must Sancho do?

Anonymous paradox— "Please ignore this statement."

paradoxes dealing with infinity —

$x+x^2+x^3+\ldots+x^{n+1}+\ldots=x/(1-x)$

$1+1/x+1/x^2+\ldots+1/x^2+\ldots=x/(x-1)$

$1+1/x+1/x^2+\ldots+1/x^n+x+x^2+x^3+\ldots+x^{n+1}+\ldots$

$=x/(x-1) + x/(1-x)= 0$ This relation should be true for any x, except $x\neq1$. But if x is positive, the left side would be greater than 0. How can this be?

$\{1, 2, 3, 4, \ldots, n, \ldots\}$ the counting numbers

$\{1, 4, 9, 16, \ldots, n^2, \ldots\}$ counting numbers squares

These two sets can easily be put into a 1-1 correspondence. How can there be the same, number in each set?

Cantor's paradox —Georg Cantor (1845-1918), the father of set theory, had shown that the set of subsets of a set contains more members than the set itself. Is this true for the set of all sets?

Russell's paradox —Bertrand Russell's (1872-1970) paradox deals with the idea of membership of a set. A set is either a member of itself or not a member of itself. Refer to a set which does not contain itself as a member as *regular.* For example, the set of people does not include itself as a member, since it is not a person. Refer to a set which does contain itself as a member as *irregular.* An example is the set of sets with more than say five elements. Is the set of *all* regular sets regular or irregular? If it is regular, it cannot contain itself. But it is the set of all regular sets, thus it must contain all regular sets, namely itself. If it contains itself, it is irregular. If it is irregular, it contains itself as a member, but it is supposed to contain only regular sets. Russell's paradox was devastating to Gottlob Frege, a German mathematician. He had just finished the second volume on the logical development of arithmetic. The appendix of his second volume begins with " *A scientist can hardly encounter anything more undesirable than to have the foundation collapse just as the work is finished. I was put in this position by a letter from Bertrand Russell...*"

These are only a small number of the many paradoxes that exist and are continually being created, intentionally or not.

In our everyday activities or in the process of creating and defining mathematical systems and ideas, a paradox can be both a nemesis and a learning tool. Mathematics evolves in a number of ways. A mathematician working on a problem may discover a new way of doing something or create a new mathematical system (e.g. discovery of non-Euclidean geometry). On the other hand a mathematician may be trying to use mathematics to describe or explain something in the real world and may devise new ideas (e.g. fractals, chaos theory) to explain real world phenomena. In the process paradoxes occur, intentionally or unintentionally. Generalizing without proof, dividing by 0, assuming something exists without verification, not adhering to definitions are just some of the things which have created startling paradoxes.

*Some additional famous paradoxes for your reference are— Curry's triangle paradox, De Morgan's paradox, Euler's paradox, Grelling's paradox, infinite regression paradoxes, Leibnitz paradox, map paradox, Petersburg paradox, Protagoras paradox, Newcomb's paradox, nontransitive paradoxes.

the game of nimbi

Nimbi was developed by the Danish scientist, Piet Hein, who adapted it from the ancient game of Nim.

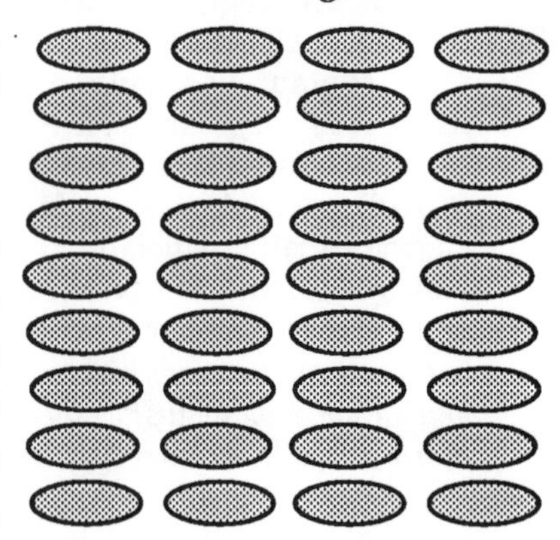

Nimbi can be played almost anywhere. All one needs are some pebbles or stones.

Procedure:

- Place as many stones as you wish in the shape of a square or a rectangle. In the diagram above, 36 stones were used in the form of a 4 by 9 rectangle.
- Each player takes as many *adjacent stones* from any row or column. For example, the first player may decide to take the first 3 stones from the 4th row, leaving:

The second player cannot take all the stones from columns 1, 2 or 3 because they each have a stone missing. But the second player may take all the stones from the 4th column because there are no gaps in the 4th column. Or the second player could take as many adjacent stones as he or she wishes from any of other rows or columns.

Object of the game:

The player who picks up the last stone is the loser.

kaleidoscopes & symmetry

The Scottish physicist, Sir David Brewster, invented the kaleidoscope in the early 1800's. It has never lost its popularity, and people continually discover and enjoy kaleidoscopes. The images that one sees in a kaleidoscope have many lines of symmetry.

A line of symmetry divides an object exactly in half so that when folded along this line, the two parts fit on top of one another exactly.

Today, sophisticated computer programs allow the user to draw designs while the computer is generating its exact mirror image — as illustrated by the above diagram.

the 7, 11, & 13 number oddity

Take any three digit whole number, abc. Write a new number by repeating its digits twice, abcabc. Any number formed in this way will be divisible by 7, 11, 13, 77, 91, 143, and 1001! How does this work?

762

762,762

762,762÷7=108,966

762,762÷11=69,342

762,762÷13=58,674

762,762÷77=9906

762,762÷91=8382

762,762÷143=5334

762,762÷1001=762

For an explanation, see the appendix.

paper model of the Klein bottle

Christian Felix Klein, a German mathematician (1849-1928), devised the topological Klein bottle. The bottle has only one surface, and it passes through itself. If water is poured into it, the water just comes out the same hole.

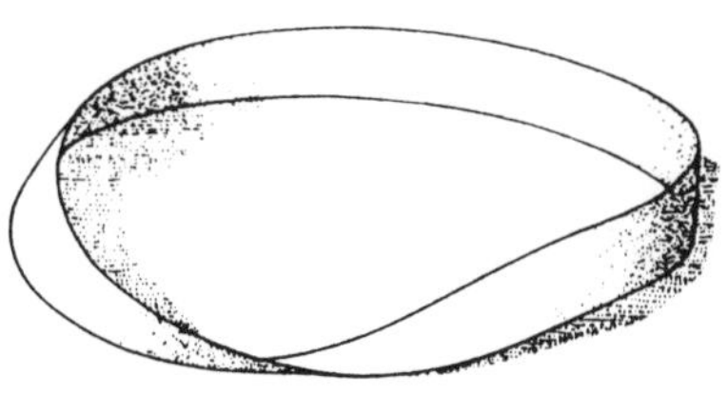

The Möbius strip was created by Augustus Moebius (1790-1868)

There is an interesting connection between the Klein bottle and the Möbius strip. If the Klein bottle is cut in half along its length, it forms two Möbius strips!

The model below is analogous to the Klein bottle with only one surface (one side). Consider cutting this model in half and see if it produces two Möbius strips.

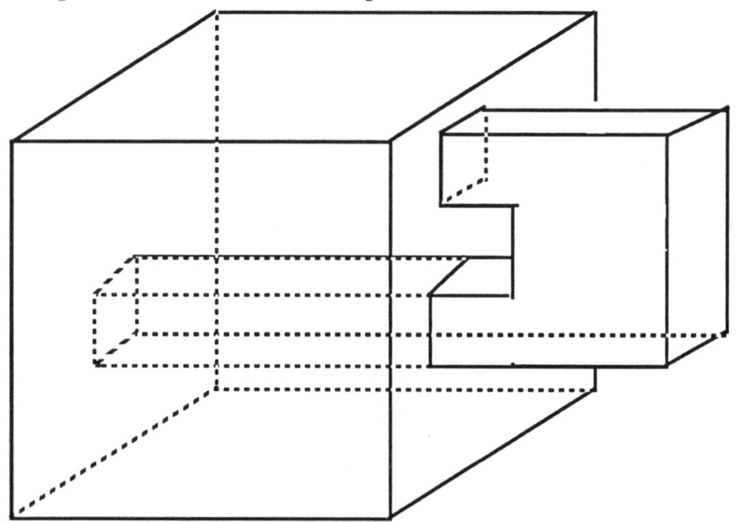

mathematical problems & discoveries

Throughout the evolution of mathematics, problems have acted as catalysts in the discovery and development of mathematical ideas. In fact, the history of mathematics can probably be traced by studying the problems that mathematicians have tried to solve over the centuries. It is almost disheartening when an old problem is finally solved, for it will no longer be around to challenge and stimulate mathematical thought.

Some of the most exciting mathematical discoveries have resulted in spin offs from mathematicians' efforts in attempting to solve "unsolved" mathematical problems or trying to prove or disprove mathematical ideas. *The three construction problems of antiquity* are probably some of the earliest mathematical challenges that led to many new discoveries—

• *the conchoid of Nicomedes for trisecting an angle*
• *the cissoid of Diocles, Archytas' method for finding the cube root of 1/2* • *Plato's cube duplicator* • *Eudoxus' kampyle* • *Eratosthenes' sliding square* • *Apollonius of Perga method for duplicating a cube by inserting two geometric means between a given pair of lines* • *Menaechus' method using parabolas for duplicating a cube* • *Sestine's method using a hyperbola and a parabola for duplicating a cube* • *the lunes of Hippocrates for squaring a circle* • *quadratrix of Hippias for squaring a circle* • *Archimedean trisection device* • *Etienne Pascal limaçon for trisecting an angle* • *gnomon for trisecting an angle* • *spiral of Archimedes for trisecting an angle* • *helix for trisecting an angle* • *Pappus' method for trisecting an angle using a circle and a hyperbola*

— and the list goes on. Over a span of more than 2000 years mathematicians in *efforts to prove Euclid's Fifth Postulate,* eventually discovered and created non-Euclidean geometries in the 19th century. Euler's solution to *the Königsberg Bridge Problem* (1736) initiated the study of topology. The *four color map problem* (first stated formally in 1852 by English mathematician Francis Guthrie) has been "proved" many times over the years. But the proofs have been flawed. Interest in the problem has not subsided, especially as the field of topology expanded. Teachers tantalized classes with it, and mathematicians have expended years trying to solve it. *The four color map problem* was finally set to rest in 1976 with a computer proof by K. Appel and W. Haken of the University of Illinois.

A woodcut from Gregor Reisch, 1503. Contest between the modern algorist (inaccurately represented by Boethius) and abacist (portrayed by Pythagoras). The goddess Arithmetica is in the background.

Recent years have brought some exciting proofs to some bewildering math problems and/or conjectures—

> *Mordell Conjecture* of the 1922 claimed that a certain class of polynomial equations only have a finite number of rational solutions (those curves associated with topological surfaces that have two or more holes). This was proven by Gerd Faltings in 1983.
>
> *Poincaré's 4-dimensional conjecture* stumped mathematicians for more than 80 years until the work of M. Freedman of U.S.C. and S. Donaldson of Oxford University.

One of the most recent and famous unsolved *problems* of mathematics is *Fermat's Last Theorem**. Pierre de Fermat, (1601-1665) a lawyer by profession, enjoyed spending his leisure time studying mathematics. He is regarded as one of the great French mathematicians, having made contributions to so many areas of mathematics, especially number theory. Although he published little of the mathematics he wrote, he corresponded with leading mathematicians of his time; and undoubtedly influenced their work. His over three thousand mathematical papers and notes have come to light via his mathematical correspondence and in notes he wrote in his translation of *Arithmetica* by Diophantus. It is in this book we find the famous note Fermat wrote in the margin—

> *To divide a cube into two cubes, a fourth power, or in general any power whatever above the second, into two powers of the same denomination, is impossible, and I have assuredly found an admirable proof of this, but the margin is too narrow to hold it.*

restated:

If $n > 2$, there are no whole numbers a, b, c such that $a^n + b^n = c^n$.

Naturally when this note was discovered after his death, the challenge was out to mathematicians. But for centuries the proof or disproof has eluded even the most prominent mathematicians. Some feel he never had a proof, but intended to frustrate some of his colleagues. Nevertheless, it stimulated an abundance of important mathematical ideas and discoveries over the past 350 years.

measuring old sayings with different units

Can you convert these sayings to their original measurements?

"Give him 2.5 centimeters and he'll take 1.6 kilometers."

"He demanded a .45 kilogram of flesh."

"A miss is as good as 5280 feet."

"The perfect yard."

"Every 2.5 centimeter a king."

"157 centimeters, eyes of blue..."

"I love you 35.2 liters and .88 decaliters."

"28.4 grams of prevention are worth .4536 kilograms of cure."

"First down and 30 feet to go."

For conversions, see the solutions section in the appendix.

mathematics & crystals

Although crystals seem to be in the limelight of "New Age", the use of crystals for healing and energizing dates back thousands of years. Today we find crystals in many objects with which we come in contact in our daily lives be it a cup, a glass, a spoon, a radio, a watch, or one's hair.

Secrets of the "power" of crystals lie in understanding their structure and growth, and applying modern technology to discover and launch their vast uses. Crystals are very ordered elements. Their structure is technically so precise, that mathematics is the perfect tool for analyzing, identifying and categorizing them.

Polyhedra, symmetries, tessellations, interfacial angles, geometry, projections, sinusoidal functions are but a few of the mathematical ideas used in analyzing crystals. By looking at the formation and use of crystals, these ideas become evident.

Every crystal can be shown to be formed from atoms which occur in one of *six* possible unit cell shapes or building blocks. These building blocks are polyhedra (a solid whose faces are polygons). The crystal polyhedra are:

1) ***Isometric system or cubic system***— *each unit is composed of congruent cubes. The faces of this polyhedron are congruent squares. Any three concurrent edges are at right angles to each other. Examples are pyrite, alum, garnet, galena.*

2) ***Tetragonal system***— *the faces of this polyhedron unit are rectangles in which any three concurrent edges are at right angles*

to each other and 2 of the 3 edges are congruent. Examples are cassiterite, rutile.

3) ***Orthorhombic system***— *three concurrent edges are at right angles to each other and none are congruent. Examples are topaz, celestile, chalcotite.*

4) ***Monoclinic system***— *2 of the 3 concurrent edges are at right angles to each other and none are congruent. Examples are borax, azurite, muscovite.*

5) ***Triclinic system***— *no concurrent edges are at right angles, nor are they congruent. Few crystallize.*

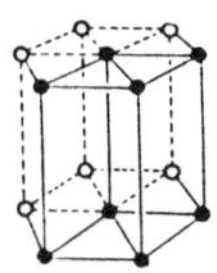

6) ***Hexagonal system***—*2 of the concurrent edges are congruent and make 60°and 120°angles to each other. The 3rd edge is at right angles to these two edges and of different length. Examples are calcite, tourmaline, beryl.*

These six systems are the building blocks for all crystals. Let's look at the crystallization (formation) in an isometric system to understand how different crystals can form from the same building blocks. One can consider the crystallization process as a 3-D form of tessellation. Isometric crystallization begins with a specific cube and generates thousands of duplicates. Then they are stacked together, and end up in the form illustrated.

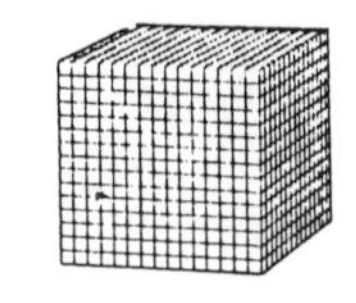

But not all isometric crystals end up looking like this. If on the other hand we remove a certain number from each corner, in other words we truncate the cube, we get instead a crystal of the shape illustrated.

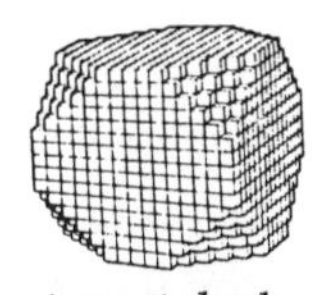

truncated cube

If we continue removing more building blocks we could end up with a crystal in the shape of a octahedron.

octahedron crystal

These examples are all isometric crystals because the basic building unit cell is a cube. Each of the six crystal systems can have thousands of different types of crystals formed depending on what is removed or added and what type of atoms comprise the unit cell. (For example, in an average grain of salt there are 5.6×10^{18} unit cells.)

Besides classifying crystals by identifying one of the six polyhedra building blocks, analyzing the symmetries they possess is equally important. Crystals can have a single type of symmetry, no symmetry or a combination of them, namely — point symmetry, line symmetry, plane symmetry.

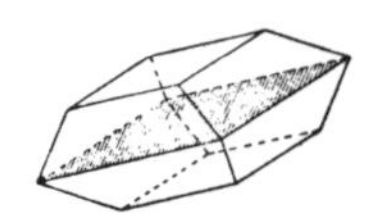
example of plane symmetry

There are different methods for determining a crystal's symmetry. With precision instruments, the angle measurements between faces (interfacial angles) are taken which help identify the symmetries a crystal has and also determine to which of the six crystal systems it belongs. Each crystal system has a specific number and type of symmetry. For example, the cubic (or isometric) system has 9 planes of symmetry, a center point of symmetry, and 13 different line (or axis) symmetries, while the triclinic system has no line of symmetries. Other means for identifying crystals are studying the geometric patterns formed when X-rays are passed through them or studying projections of lines (lines perpendicular to their faces) on a sphere that surrounds a particular crystal.

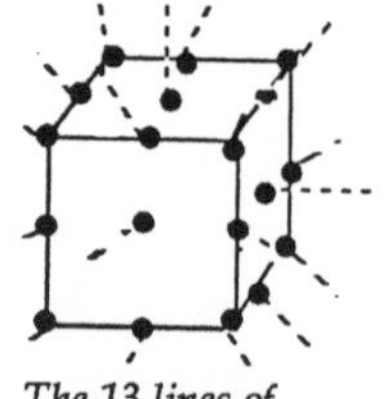
The 13 lines of symmetry of a cube.

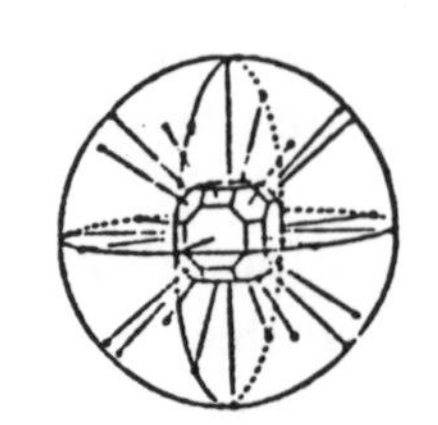
a spherical projection

Not until the end of the 19th century was it discovered that quartz crystal generates electrical charges when pressure is applied to it. Conversely when electrical current was passed through the quartz, the quartz regulated the current through its uniform vibrations. In other words, quartz crystal works in both directions—creating electricity when pressure (in all sorts of forms) is applied and regulating the electricity. This two-way property is called piezoelectricity. Knowledge of quartz crystal's structure and its properties was essential to launch this power. The quartz had to be either mined and cut to exact specifications or, as is done today, artificially grown. Precision is imperative, otherwise a crystal with an incorrect frequency could cause havoc. Some examples are: (1) a radio broadcasting station's signal may be picked up on the wrong radio frequency number or (2) a quartz crystal with the wrong frequency would interfere rather than manage the flow of information in a computer.

Our knowledge of crystals and their uses is still at the evolving and discovery stages. The following is a partial list of areas where quartz crystals are used or may be used in the future—

- *watches, computers, VCRs, washing machines, pilotless stoves, microwave stoves, dishwashers, new automobile engines*
- *converting sound waves to electrical signals —sonar*
- *monitoring broadcasting frequency*

possible future uses

- *warning devices for impending earthquakes, structural imbalance of a building or bridge by translating vibrations to electrical signals*
- *convert ocean wave action into usable energy*
- *voice and pen recognition devices for personal identification*

Research has thus far focused on the quartz crystal. The study and discoveries of other properties of crystals is yet to come. Mathematics will be an invaluable tool for research and for enhancing our understanding of these exciting minerals.

Chinese rod numerals

Many of us are familiar with various methods and means of recording numerations, such as the quipu of the Incas , the symbols | and ∩ of the Egyptians, the 𒁹 and 𒌋 of the Babylonians, the alphabet of the Greeks. However, the ingenious Chinese rods numerals are less familiar.

The first record of Chinese numerals —

appeared as notches on divination bones and tortoise shells[1] during the Shang dynasty around 14th century B.C.. From these symbols the Chinese evolved two methods to write numbers, both of which were base ten positional systems.

One method used the written characters of the words for the numbers. For example,

one	*two*	*three*	*four*	*five*	*six*	*seven*	*eight*	*nine*
一	二	三	四	五	六	七	八	九

are the characters for these Chinese words. Naturally there are variations of the styles and forms depending on whether the characters were used in formal or informal documents.

The other method used numerals developed mainly for calculating which were called "rod numerals". Between the 2nd century B.C. and the 2nd century A.D. rod numerals were formed into a sophisticated method of numeration. Initially, rods or bamboo sticks were used to form the numerals. Later, counting boards were made to hold the rods and a blank space was used to designate the absence of units for that place value.

The rods were laid out in one of two ways to represent the digits from 1 to 9.

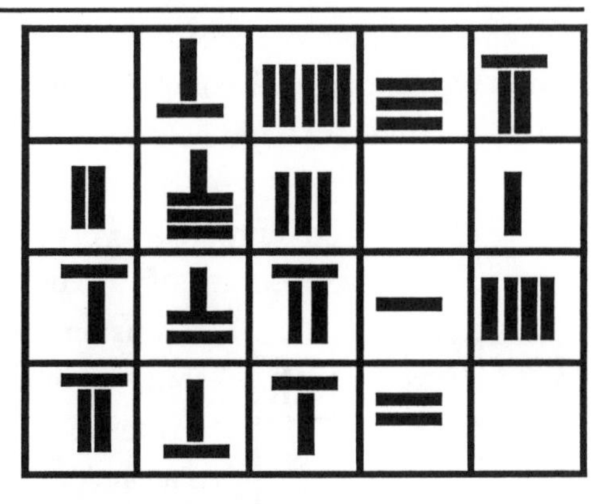

This counting board illustrates rod numerals for the following numbers (top to bottom): 6537; 28301; 67714; and 76620. The Chinese also used the counting boards to set up systems of equations. Here the top row could also be used to represent the equation $6w+5x+3y+7z=0$.

row 1
row 2
1 2 3 4 5 6 7 8 9

Two possibilities were developed to avoid mistaking a number such as 12 for a 3. If only one set of digits were used the numeral for 12 could easily be confused with that of 3. For example — 12 would be | || and 3 would be |||. To avoid this mix-up, the Chinese alternated the digits from the 1st row with those from the 2nd row. It became customary to use the numerals from row 2 for odd place value powers of 10, and the rod numerals from row 1 for the even and 0 powers of 10 (10^0, 10^2, 10^4, an so forth). To write 12 the rods would appear as = |

Therefore, the number 2816 would be formed in the following manner

Thus, the following numbers would be written as:

4   31   132   5682

To indicate a zero placeholder in a number, such as 205, a space had to be left. Probably to keep the rods from moving or rolling, counting boards were later developed. Counting boards were subdivided into a grid of little boxes, which held the rods for a particular digit of a number. These boards stabilized the rods, and also clarified the location of any zero placeholder. The boards were carried for computation work, and functioned a lot like today's modern hand held calculators.

二百七
207

Eventually the rod numerals evolved to their written form. But the absence of a zero symbol caused confusion in the writing of some numbers. Initially, when a written rod numeral needed a zero placeholder, e.g. 207, it was either written in the classic characters or grid marks were drawn showing the location of an empty space, as if the number were done with a counting board.

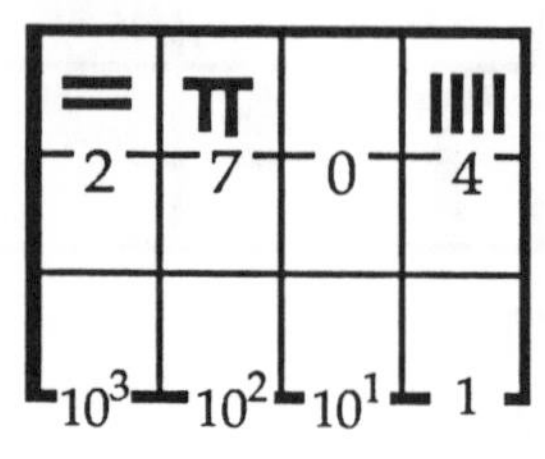

It was not until the 8th century A.D. that the symbol O was adopted for the zero[2]. The Chinese used this symbol both as a placeholder zero and as a decimal point. With the introduction of decimal numbers for fractions Chinese computation and arithmetic methods flourished.

The Chinese system of numeration, which had evolved independently from systems of other cultures, was a base ten place value system from its inception.

The decimal .036 would be written this way. The first O represented the decimal.

[1]This may explain the origin of the legend of Lo shu appearing on a tortoise shell.

[2]Many believe that this was a result of the Chinese contact with Indian mathematicians during this time period.

tethered goat puzzle

A goat is enclosed in a 2 π acre fenced area in the shape of an equilateral triangle. It's tethered to a post at one of the triangle's vertices. To the nearest foot, determine the length of a rope which will allow the goat to eat only one-half of the grass in these 2π acres.*

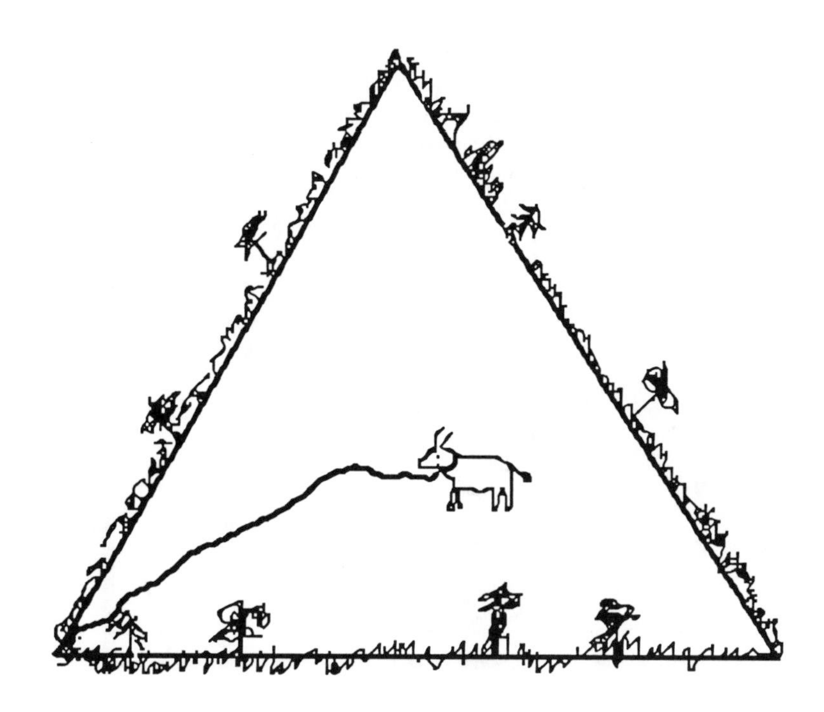

For solution, see the appendix.

* This puzzle was adapted from one by renowned English puzzlist Henry Dudeney (1847-1930)

Sam Loyd's hidden 5-pointed star puzzle

One of America's greatest and most famous puzzlist was Samuel Loyd (1841-1911). His career started in his teens with his prize winning chess problems. At the age of sixteen he became

Find the concealed five pointed star.

problem editor for the *Chess Monthly,* and in later years he edited other chess columns for various newspapers and magazines, including the *Scientific American Supplement.* But he is most famous for his ingenious puzzles, games and novelty inventions. These include his *Trick Donkeys, 14-15 puzzle* (slide puzzle, whose popularity continues today), his adaptation of *Parcheesi,* and his famous *Get off the Earth puzzle.* The *hidden five-pointed star puzzle* is from his book, *Cyclopedia,* which was compiled by his son after his death.

Egyptian hieratic numerals

Egyptians used hieroglyphic script and numerals to inscribe their monuments, buildings and stone tablets. The scribes, in order to save time and write more quickly and efficiently when using reeds and papyrus, gradually changed and simplified the hieroglyphic script and numerals to their hieratic form. Egyptian scribes slowly adapted hieroglyphic notation to hieratic numerals for accounting, computation, and general record keeping. Eventually Egyptian hieroglyphic notation was used only for decorative purposes. From the third to the first millennium B.C., the hieratic script was used in all areas of work, such as scientific, legal, administrative, religious, literary. In the 12th century B.C., hieratic script was replaced by another cursive script called demotic.

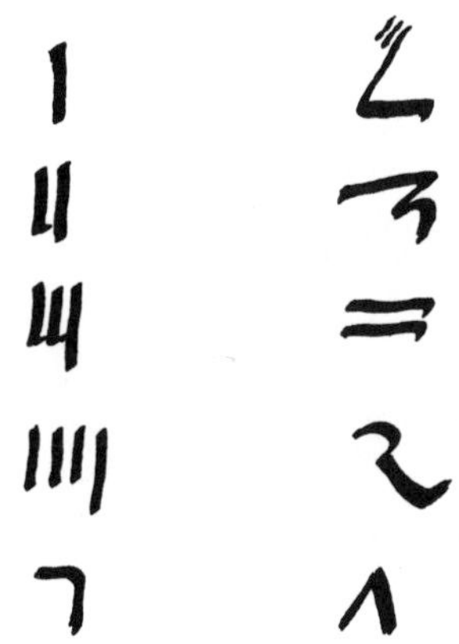

The hieratic numerals from 1 to 10

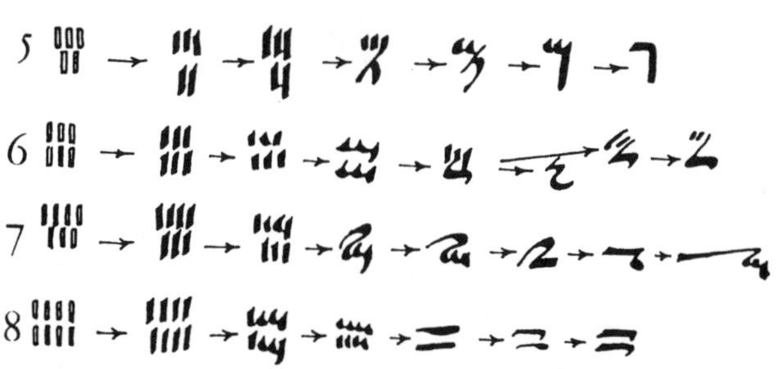

5

6

7

8

This diagram shows the changes that some of the hieroglyphic numerals underwent.

calendars & time measurement

Calendars are the giant clocks of the universe. Our methods for keeping track of minute periods of time have been refined with astounding accuracy. Over the centuries our time pieces have been revised and their accuracy improved. We have progressed from the sundial to the water clock, the pendulum clock, the mainspring clock to today's quartz watch. But the progression and improvement of the calendar has thus far stopped since the introduction of the Gregorian Calendar in 1582.

This calendar was fashioned out of beaten gold, and used in Peru between 800 B.C. and 200 A.D.

History

The inhabitants of Mesopotamia, first the Sumerians and then the Babylonians, are credited with the first calendar. In the 14th century B.C., the Chinese established the year as 365 1/4 days and the lunar month as 29 1/2 days. The works of Hesiod and Homer established the Greeks as using lunar calendars in the

13th century B.C. In the late 10th century there is reference to the Hebrew lunar calendar and the lunar-solar Hindu calendar. But the Mayan calendar was the most advanced. It was a calendar that kept track of various astronomical cycles — of days (24 hours of the Earth's rotation), of lunar months, and of the solar year (the length of time for the Earth to orbit around the Sun), in addition to other astronomical cycles, such as for the planet Venus.

In different regions of the world, the calendars that were developed reflected the seasons— such as the three seasons of the Egyptian calendar, the two seasons of the Babylonian, the four seasons of the Greek.

Today, the Gregorian calendar is used in the western world and in the business world, but we also find the Hebrew, Islamic, Hindu, Chinese, Balinese, New Guinean calendars — specializing in the needs, cultures, and religions of these people.

Prior to the adoption of the Gregorian calendar, the Julian calendar was used, which gained one year every 128 years. The solar year is 365 days 5 hours 48 minutes 46 seconds or 365. 24219074164. The Gregorian calendar was set up to have 365.2425 days per year, and eliminated any leap years that landed on a century year not divisible by 400 (e.g. 1700, 1800, 1900). These reforms lessen the calendar's discrepancy to approximately 1/4 minute every year, which adds up to one day in every 3323 years. It is a definite improvement over the Julian calendar, but it is not perfect. In 1930, the World Calendar Association developed a very systematized calendar, and it has even been discussed in the UN. But no change is foreseen in the near future. Can a perfect calendar be devised with the units of time (seconds, minutes, hours, days, weeks, months) we use? Or do we need to devise new units, e.g. a unit of time set to the length of an arc π units with six π months in a year or devise some other unit that can synchronize the various time measurements? Perhaps we need to expand our calendar, and not just have it measure the passage of time for the Earth, but instead in relation to the Universe. Space exploration is expanding our world, and a new calendar may be necessary for this aspect of our lives. It seems that most people are satisfied with our current calendar, and would be loathe to change.

JANUARY

S	M	T	W	T	F	S
1	2	3	4	5	6	7
8	9	10	11	12	13	14
15	16	17	18	19	20	21
22	23	24	25	26	27	28
29	30	31				

FEBRUARY

S	M	T	W	T	F	S
			1	2	3	4
5	6	7	8	9	10	11
12	13	14	15	16	17	18
19	20	21	22	23	24	25
26	27	28	29	30		

MARCH

S	M	T	W	T	F	S
					1	2
3	4	5	6	7	8	9
10	11	12	13	14	15	16
17	18	19	20	21	22	23
24	25	26	27	28	29	30

APRIL

S	M	T	W	T	F	S
1	2	3	4	5	6	7
8	9	10	11	12	13	14
15	16	17	18	19	20	21
22	23	24	25	26	27	28
29	30	31				

MAY

S	M	T	W	T	F	S
			1	2	3	4
5	6	7	8	9	10	11
12	13	14	15	16	17	18
19	20	21	22	23	24	25
26	27	28	29	30		

JUNE

S	M	T	W	T	F	S	
					1	2	
3	4	5	6	7	8	9	
10	11	12	13	14	15	16	
17	18	19	20	21	22	23	
24	25	26	27	28	29	30	W

JULY

S	M	T	W	T	F	S
1	2	3	4	5	6	7
8	9	10	11	12	13	14
15	16	17	18	19	20	21
22	23	24	25	26	27	28
29	30	31				

AUGUST

S	M	T	W	T	F	S
			1	2	3	4
5	6	7	8	9	10	11
12	13	14	15	16	17	18
19	20	21	22	23	24	25
26	27	28	29	30		

SEPTEMBER

S	M	T	W	T	F	S
					1	2
3	4	5	6	7	8	9
10	11	12	13	14	15	16
17	18	19	20	21	22	23
24	25	26	27	28	29	30

OCTOBER

S	M	T	W	T	F	S
1	2	3	4	5	6	7
8	9	10	11	12	13	14
15	16	17	18	19	20	21
22	23	24	25	26	27	28
29	30	31				

NOVEMBER

S	M	T	W	T	F	S
			1	2	3	4
5	6	7	8	9	10	11
12	13	14	15	16	17	18
19	20	21	22	23	24	25
26	27	28	29	30		

DECEMBER

S	M	T	W	T	F	S	
					1	2	
3	4	5	6	7	8	9	
10	11	12	13	14	15	16	
17	18	19	20	21	22	23	
24	25	26	27	28	29	30	W

For the WORLDSDAY CALENDAR, each column of months always starts on the same day. All holidays would be on the same day each year, except Easter. Note a Worldsday is placed between December and January. For every leap year, its extra day (another Worldsday) is placed between June and July.

changing the day

Dr. Dennis McCarthy of the United States Naval Observatory reported that on January 24, 1990 the Earth's day was lengthened by 5/10,000 of a second. Changes in the speed of the rotation of the Earth are calculated by measuring movements of stars and planets. Westerly bursts of wind coming from Asia across the Pacific Ocean caused the change. The Earth's rotation can be altered minutely by the effects of weather from the forces of winds on the planet's surface. During the time period of 1982-1983, the Earth slowed by 2/10,000 of a second. This was due the 1982-1983 El Niño* which caused changes in the ocean water temperature and and air pressure which affected world wide weather conditions. This El Niño was considered the most severe occurrence of its type in the past 100 years.

---

* An El Niño is a warming of the equatorial Pacific Ocean, and linked to the Southern Oscillation, which is a periodic change in large-scale Pacific Ocean weather systems.

space filling curve & population

In a space filling curve, every point within a given area or in a 3-D space is traced and gradually blackens the space. The manner in which a space filling curve expands and generates itself seems analogous to population growth. The Peano space filling curve demonstrates how this curve can grow and continually expand itself within the confines of a given area or space. The population of a city, or more generally the world, is continually growing, but is confined to fixed boundaries. Consequently, fractals, space-filling curves and computers using available data can be used to project population densities in different regions.

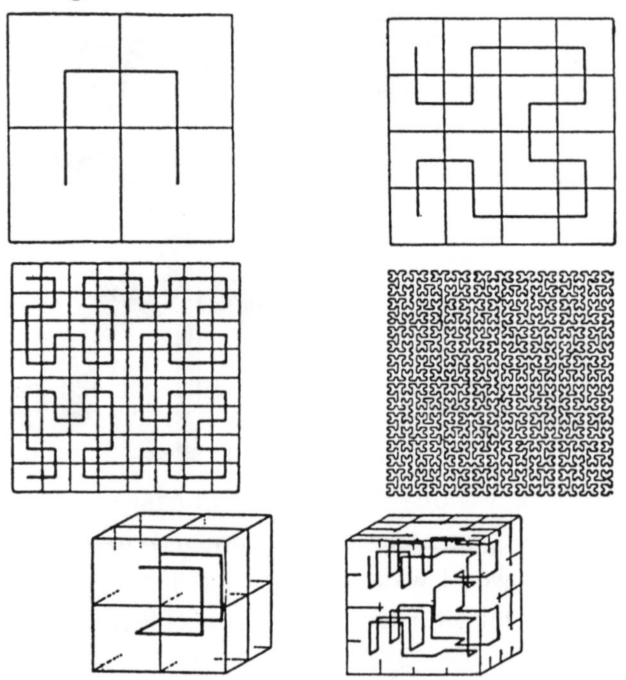

This example shows the stages of a space-filling curve which envelopes the space of an entire cube by continually generating itself in the special way illustrated.

convergence/ divergence optical illusion

In mathematics we can never only rely on our eyes for measurement. No matter how acute our vision is, the physical structure of our eyes has limitations and our mind can play tricks on us. For the illustration below, the top figure appears smaller, yet if you were to cut it out and slide it below the bottom figure, it would appear larger. This

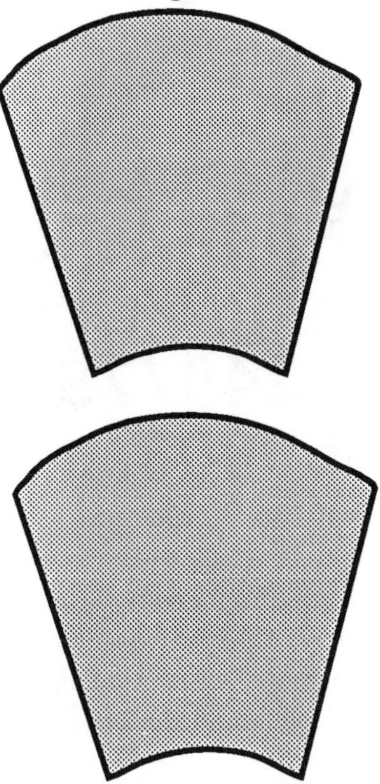

illusion is called convergence/divergence—created by angles or segments which lead our eyes inward or outward, thus shortening or lengthening an object. Artists, architects, and designers of clothing realize the power of this illusion, and incorporate it in their works.

e and banking

What does **e** have to do with everyday matters? In fact, it is used in daily routines just as much if not more than any particular whole number, even though we may not always be aware of its presence. Few people know that **e** is actually a number. Ask people what **e** is and most will say, it's the 5th letter of the English alphabet. Some will know it's a "strange" number they came across in a math course. Fewer will know it as an irrational and a transcendental number.

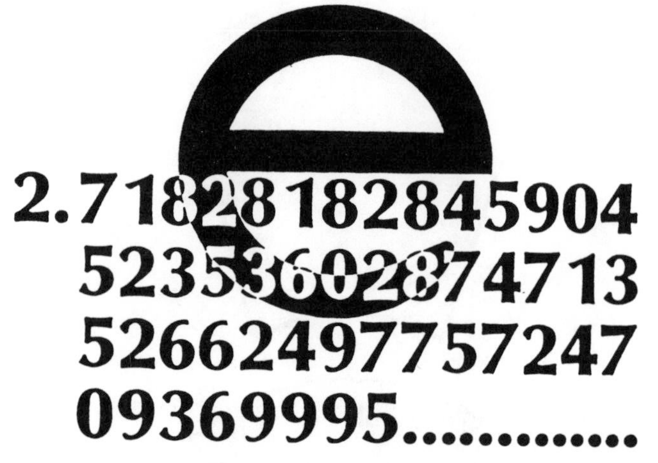

In banking today, it is the one number most helpful to bankers. One may ask how a number such as **e** be in anyway related to banking, which deals specifically with dollars and cents. If it were not for the discovery of **e**, bankers would have had an horrendous time computing today's interest rates, either those compounded daily or those compounded continually. Fortunately **e** comes to the rescue.

The definition of **e** is the limit of the sequence $a_n=\{1+1/n\}^n$, which is usually written $\lim_{n\to\infty}(1+1/n)^n$. How does this formula help in computing interest rates? The actual formula for calculating an interest rate plus accumulated principal is A=P

$(1+r/n)^{nt}$, where A=accumulated money, P=money (principal) you started with, r=interest rate, n=number of times the rate is compounded annually, t=the number of years the money is in the savings account. This formula transforms to the formula for **e.** When one invests $1 at 100% interest rate for 1 year which is compounded indefinitely, then the amount accumulated would be **e dollars.** At first thought, one may think the amount would be astronomical, but approximating the value of **e** from this formula gives the following estimate for **e.**

$(1+1/n)^n$:	n=1	n=2	n=3	n=100	n=1000
	$(1+1/1)^1$	$(1+1/2)^2$	$(1+1/3)^3$...	$(1+1/100)^{100}$...	$(1+1/1000)^{1000}$
	2	2.25	2.370...	2.704813...	2.716923....

So we see to the nearest penny, our investment of $1 would never go beyond $2.72. In fact, **e** =2.7182818284590452353602 when carried out to twenty-two decimal places.

The next question is how does the formula $A=P(1+r/n)^{nt}$ work? The best way to determine this is to experiment. Say we start with $1000 and deposit it in a bank for 1 year at 8% interest. Let's see what happens when it is first compounded annually, then semiannually, and then quarterly.

annually:

$1000(1+8\%/1)^{1(1)}$

=1000 + 80(from interest)

=1080

semiannually:

$1000(1+8\%/2)^{2(1)}$

$=1000(1+.04)^2$

=1000(1+1.04)(1+1.04)

P_1,

principal after the first

$= P_1(1+.04)$

$= P_1 + P_1(.04)$

1st half 2nd half

quarterly

$1000(1+8\%/4)^{4(1)}$

$=1000(1+.02)^4$

=1000(1+.02)(1+.02)(1+.02)(1+.02)

P_1 (1+.02)(1+.02)(1+.02)

P_2(1+.02)(1+.02)

P_3 (1+.02)

P_4

Thus, if compounded daily one computes $1000(1+8\%/365)^{365(1)}$. This would take a fair amount of time to calculate by hand. But with modern day electronic calculators and especially computers the result is arrived at in an instant.

domino puzzles & more

A domino is formed by joining **two congruent squares** together.

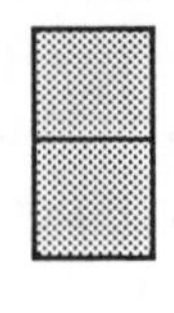

The Domino Puzzle

Which of these diagrams can be covered using only dominoes? Remember the dominoes must be the same size and cannot be placed on top of each other.

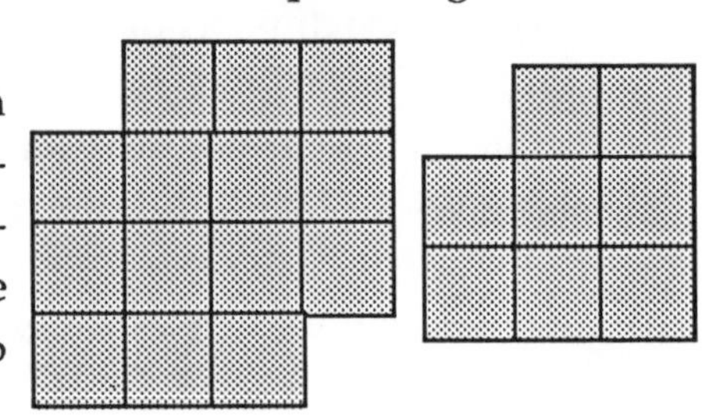

The Tromino Puzzles

Trominoes are formed by joining three congruent squares together. Here are two possible tromino shapes.

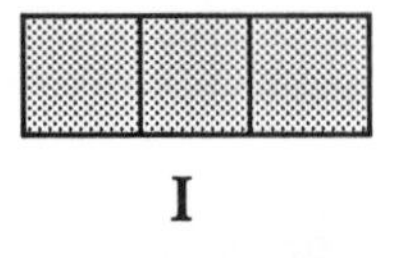

I

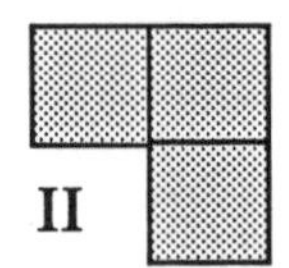

II

Can this shape be covered using only type **I** trominoes? type **II** trominoes?

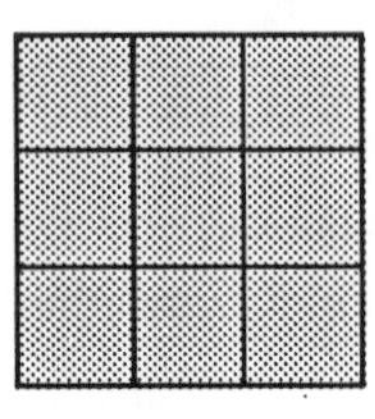

Tetromino Puzzles

Tetrominoes are made from four congruent squares. Their possible shapes are—

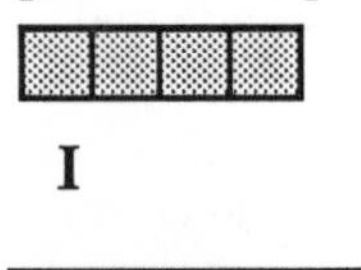

I

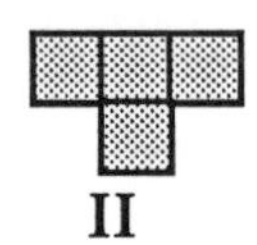

II

III

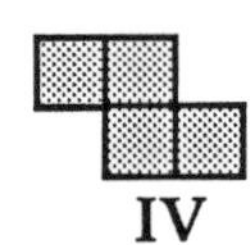

IV

?

V

Draw a fifth tetromino?

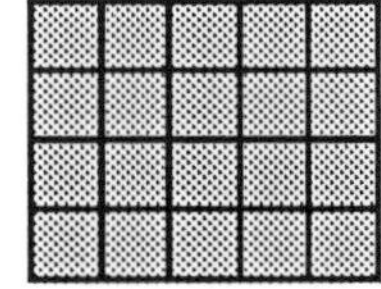

Can this shape be covered with type **I** only? type **II** only? type **III** only? type **IV** only? type **V** only?

Don't stop with tetrominoes. Make up your own puzzles with pentominoes, hexominoes, heptominoes, ...

physically Pythagoras

Paperfolding has been used to prove many geometric theorems, including the one stating that three angles of any triangle total 180º. Here is a paper folding proof of the Pythagorean theorem.

Given a right triangle, prove
$$a^2 + b^2 = c^2$$

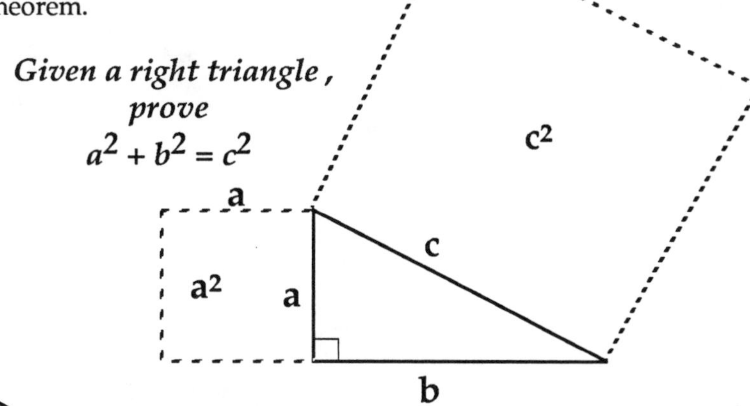

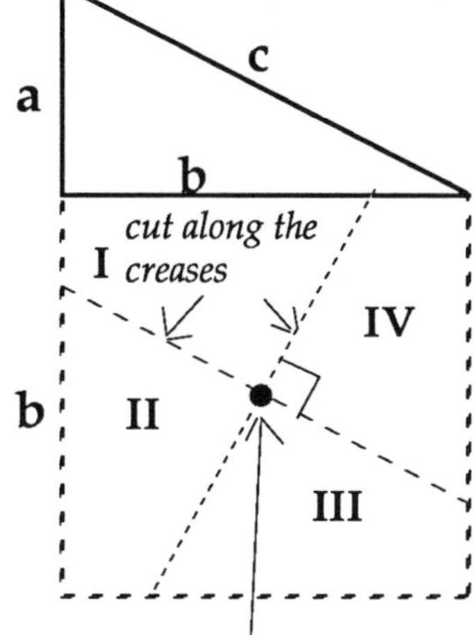

Find the square's center by finding the intersection of the square's diagonals.

Rearrange the five pieces from b's square and a's square as shown below. They form side c's square, thus showing $a^2 + b^2 = c^2$

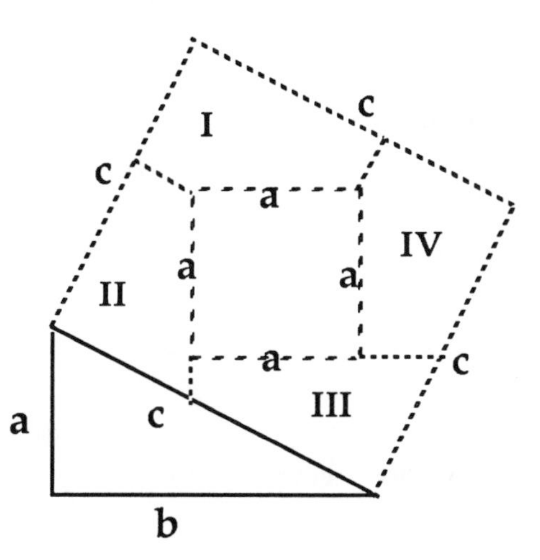

yo-yo math

There is more to yo-yoing than meets the eye. One might think that working a yo-yo is like tossing a rock. But no, the physics behind yo-yoing resembles that of the rolling ball. The yo-yo's string acts as its plane of incline. If the yo-yo were just dropped it would arrive on the ground more quickly than spinning it on the string because as the string unwinds the "axle" gets smaller. The yo-yo spins more slowly the larger the diameter of its axle. Consequently it spins more slowly at the beginning of its descent, and gains momentum as the string unwinds from around its axle. But since it spins more slowly in the beginning more energy is available for falling and its velocity is greater in the beginning of its trip. At the bottom of the string the yo-yo is now spinning more quickly and less energy is available. In fact, the yo-yo begins to lose speed midway down.

The first yo-yo had the string fixed to the axle so the yo-yo always had to go up and down. But then the yo-yo was modified by Donald Duncan, the father of the modern yo-yo. The string was looped around the axle, enabling the yo-yo to spin on the loop around its axle. To allow the yo-yo to free spin around its loop and minimize the loss of spin, the string must be kept from hitting the sides of the yo-yo. To this end designers have made a new high-tech yo-yo with a concave axle coated with Teflon. Innovations will probably not stop here.

creating mathematical mosaics

With the accessibility of computer graphics, mathematical mosaics (*tessellations)* have taken on a whole new perspective. Programs are now available in which the click of a mouse allows a design to be effortlessly translated, rotated, and/or reflected exactly. Can you imagine how such a tool would have affected the works of Moslem artists/mathematicians at the *Alhambra* or of M.C. Escher. The knowledge of the mathematical basis of tessellations will help the artist in the design of his or her creations. Perhaps the computer age will introduce a whole new set of tessellations.

Here are some ideas that were discovered over the centuries for tessellating planes and space.

•The first observations of tessellations must have been of nature's hexagons in the honeycomb. In fact the Greek mathematician Pappus of the 4th century B.C. observed that bees only used regular hexagons to make their honeycomb cells, and that the hexagon needs the least material while forming the largest space.

•The Greeks discovered and proved that the equilateral triangle, the square and the hexagon are the only three regular polygons that tessellate a plane. Many of their tiled floors from ancient times illustrate the artisans use of regular polygons.

•A combination of regular polygons can be used to create a mosaic pattern, but one must be certain that the angles of polygonal tiles joined at the vertices total 360 degrees. Otherwise gaps or overlapping tiles will occur.

no boundary tic-tac-toe

Discovering strategies for games requires logic. Logic and intuition are at the foundation of mathematics. Unlike the conventional game of Tic-Tac-Toe, where the two players are confined to nine spaces for the playing area, *No boundary Tic-Tac-Toe* uses as many square spaces as the game needs. There is never a draw unless both players are exhausted and mutually agree to give up. Graph paper is ideal for playing the game. The game can be made longer or shorter by deciding *how many in a line make a winner.* Start with *three in a line* for a beginning game. Then go to *four in a line,* and for a longer game go for *five* or *six in a line.*

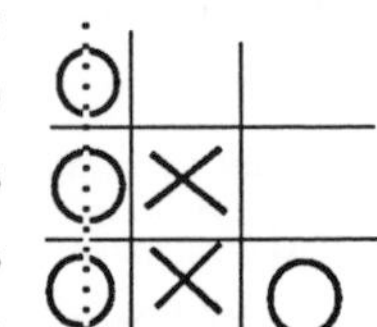

mathematical hoaxes

Mathematicians are not immune to playing pranks. Because of the seriousness with which people take the subject of mathematics, they are often caught off guard and when the prank is revealed, some fail to see the humor. Here are some famous mathematical hoaxes—

Fermat's last theorem[1]*:* As we know this theorem remains unproven. The question is: Did Fermat really prove this theorem, or did his mathematical devilishness come into to the picture? Did he just say he had proven it, knowing full well it would definitely be a frustrating experience for his colleagues to believe he knew something they did not. If his deception was intentional, wouldn't he be delighted to know how long it has gone on and how famous it has become.

Sam Loyd's origin of the tangram[2]*:* The American puzzlist Sam Loyd must have loved to play pranks. His early ones appeared as unusual and sometimes bizarre solutions for his chess columns. One of his more intriguing hoaxes was his story of the historical origin of tangrams, as presented in his book, *The Eighth Book of Tan*. He authoritatively wrote a fictional account of how the tangram evolved, it was so convincing that readers and writers even today are deceived and perpetuate the myth in their work.

Martin Gardner's four-color problem hoax: Martin Gardner should be given an award for his *four-color map problem* hoax in the April 1975 issue of *Scientific American* April. Intending it as an April Fools' article, Gardner never thought anyone would take the article seriously. But to his astonishment, over 1000 letters were received by readers who did indeed believe the story. I must admit being a perpetuator of the Gardner hoax. In the September 1989 edition of *The Joy of Mathematics* I make reference to the 1974 W. McGregor map which claims to need five colors.

The Nicholas Bourbaki hoax: Who is Nicholas Bourbaki? Although the person Nicholas Bourbaki is non-existant, the mathematical books published under his name are very real and serious. The name Nicholas Bourbaki encompasses a group of cryptic mathematicians who have written many volumes of *Éléments de mathématique* under the Bourbaki pseudonym. The ficticious Nicholas Bourbaki is a Frenchman with a Greek name from the city Nancy and connected to the fictional University of Nancago. Although the mathematical works and ideas continue to flourish by Bourbaki, no one knows for certain all the members of this secret group.

---

[1]See page 150 for the background on his theorem.

[2]See page 6 for background information.

origins of the arithmetic triangle

Most often this triangle of numbers is associated with the mathematician Blaise Pascal. In fact, it has come to be known as Pascal's triangle. He first wrote about it (circa 1653) in a book entitled, *Traité du Triangle Arithmétique (Treatise on the Arithmetic Triangle).* But this triangle of numbers was known and written about many years prior to Pascal. The works of Chinese mathematicians Yang Hui's and Chu Shih-chieh (book *Precious Mirror*) include both the Arithmetic triangle and summations of series that appear in the triangle. In the beginning of *Precious Mirror (1303)* there is an illustration of the Arithmetic triangle. Its caption, *The Old Method of the Seven Multiplying Squares,* shows the coefficients for the binomial expansion through the eighth powers. Chu refers to the existence of the Arithmetic triangle as an old method for finding eighth and lower powers. There are other Chinese works of 1100 A.D., which show a system for tabulating binomial coefficients and imply the possible existence of the Arithmetic triangle. Evidence shows that the binomial theorem and the arithmetic triangle were also known at this time to Omar Khayyam (1505?-1123), a Persian poet, astronomer and mathematician. In his book *Algebra* , he

Eyn Newe
Unnd wolgegründte
underweysung aller Kauffmanß Rechnung in dreyen Büchern/mit schönen Regeln unñ fragstucken begriffen. Sunderlich was fortl unnd behendigkait in der Welschen Practica unñ Tolletn gebraucht wirdt/ des gleychen fürmalß weder in Teützscher noch in Welscher sprach nie gedrückt. durch Petrum Apianũ von Leyßnick/d Astronomei zu Ingolstat Ordinariũ/ verfertiger.

The title page from Rechnung (1572) by Peter Apian.

wrote that he had mentioned elsewhere his discovery of a rule for determining the 4th, 5th, 6th and higher powers of a binomial. Unfortunately, that work is not extant. The earliest surviving Arabic work containing the Arithmetic triangle is from the 15th century by Al Kashi. *Rechnung* (1527) by Peter Apian has the Arithmetic triangle printed in its title page, and *Arithmetica integra* (1544) by Michael Stifel also includes "Pascal's" triangle.

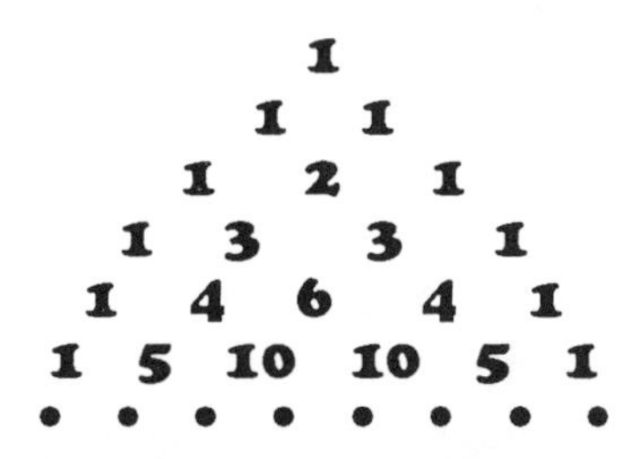

the arithmetic triangle out to six rows

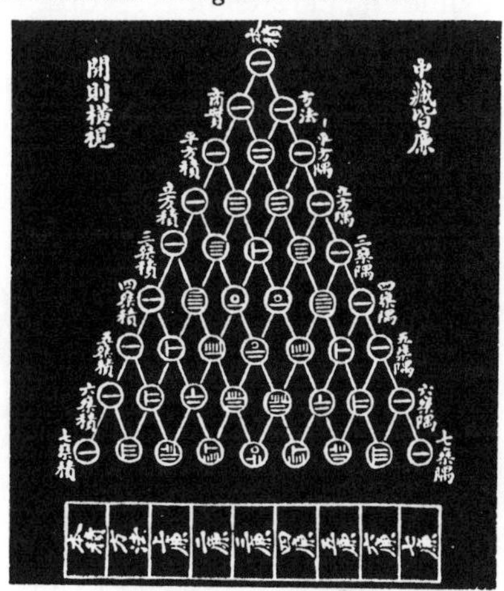
the Chinese version of the arithmetic triangle

Although Pascal was not the originator of the Arithmetic triangle, he must be recognized for discovering and proving its new properties.

redwood trees
mathematics & nature

Nature never ceases to amaze. When one takes a close look at any facet of nature's realm, we must conclude, nature knows her mathematics. The tall coast redwood and giant Sequoia trees of California are some of the oldest living things on earth. In them we can discover certain mathematical ideas, namely — concentric circles, concentric cylinders, parallel lines, probability , spirals and ratios.

• *concentric circles, cylinders and parallel lines*

At Muir Woods National Monument, a few miles north of San Francisco, one find a grove of giant redwoods. On display at Muir Woods is a cross-section of an ancient tree with historical dates recorded along its concentric rings. Among the markers are those pointing to the birth of Christ, the Norman conquest, Columbus' discovery of the New World and so on.

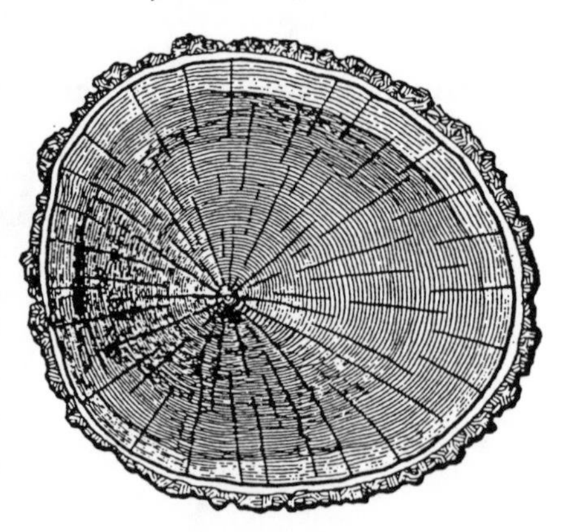

A horizontal cross-section of a tree shows concentric circles formed. A tree normally grows one circle per year, and the width of the ring depends on climate. Dry seasons produce narrower rings. Besides using these rings to approximate the age of a tree, the rings also reveal information about weather and other natural phenomena that affected its growth. Scientists can use these rings to substantiate hypotheses about such events as drought, fires, floods and famine.

When visualizing the entire length of the tree, these concentric circles represent concentric cylinders. A vertical cross- section of these cylinders forms a series of parallel lines. The parallel lines toward the center represent heartwood (dead cells). Next to these are parallel lines of sapwood which carry nourishment up and down the tree. As the tree grows, sapwood cylindrical layers become heartwood layers. Between the bark and the sapwood is a single-celled cylindrical layer called the cambium. New cells are made by the cambium which become either bark or sapwood.

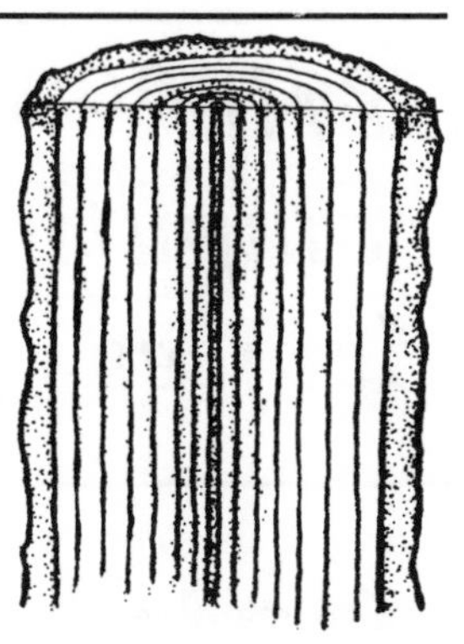

• *probability*
The sizes and quantities of seedlings vary greatly between species of trees. For example, there are 27 buckeye seeds per pound compared 120,000 redwood seeds per pound. Each cone of the redwood tree is between 1/2" and 1" long and carries 80 to 130 seeds which can remain fertile for as long as 15 years. In fact, a giant redwood produces several million seeds per year. Through numbers and sizes of seeds, nature compensates for the probability of a seed's germinating. Many small seeds increases a redwood's chances under existing adverse conditions. Even after a seed germinates the odds are several thousands to one that it will develop into a mature redwood.

• *spirals*
Glancing at the bark of the redwood tree, one notices a slight swirl in the growth pattern. Indeed, this is a growth spiral which is caused by both the revolution of the Earth and the manner in which the redwood grows to seek sunlight in the dense forest.

• *ratios*
There is an astonishing root system that supports the height of each of these giant trees. The root system is composed of mainly shallow roots (4-6 feet deep), but these roots are able to support the enormous redwood columns by branching laterally outward. This root system is usually in a ratio of 1/3 to 2/3 the height of the tree. For example, if the tree is 300 feet tall, the root system branches out between 100-200 feet from the base of the trunk to provide a firm foundation!

early computing devices

There is nothing so troublesome to mathematical practice…than multiplications, divisions, square and cubical extractions of great numbers…I began therefore to consider …how I might remove those hindrances. —John Napier

Not so many years ago the slide rule was a portable calculator for scientists and students. Today slide rules have been replaced with small programmable calculators that can instantaneously perform an entire gamut of operations from adding, subtracting, multiplying, and dividing multi-decimal figures to taking the nth root and powers to storing values of π and e, to performing metric conversions, to figuring loan amortization instantly, to even calculating and displaying graphs of functions. It is interesting to see how these high tech instruments evolved over thousands of years.

Our very first calculating device was and is our hands. As time passed, a hand number system developed for communicating among merchants and others who did not speak the same language. Even today we see young students either counting or carrying on their fingers.

The quipu was the accounting device used by the Inca. It was formed from knotted ropes and stored accounting records for the entire Inca empire.

But when numerical figures began to exceed our ten fingers, new devices had to be explored. Here enters the stacking pebbles method. But still a person had to perform the major work of grouping and counting the stones. Someone then came up with the idea of a portable pebble device which eventually gave rise to the abacus. Various

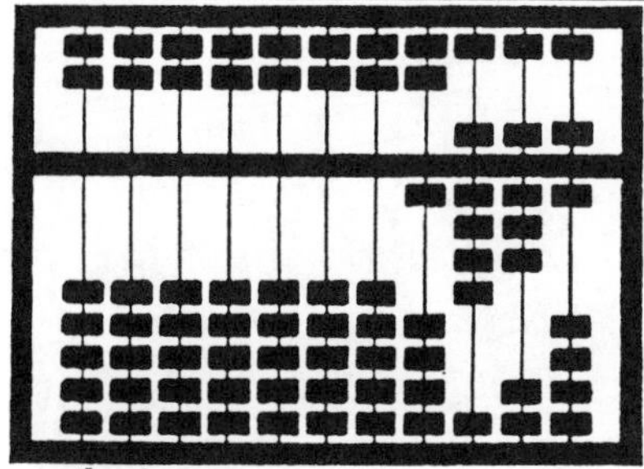

the abacus

types of abaci were used in China, ancient Greece and Rome. Abaci are still used in many parts of Asia today. The Chinese first used bamboo rods for calculating (circa 542 B.C.) and their abacus entered the computing world around the 12th century. The abacus was considered adequate for traditional calculations by merchants for centuries to follow.[1] As mathematicians tackled various problems that required either very large or very small numbers with complicated decimal work, the abacus did not suffice and manual calculations were too time consuming and prone to errors. A new method had to be devised. In the 17th century, John Napier (Scotland) invented logarithms along with a set of rods called Napier's bones. Merchants would carry a set of rods, made from ivory or wood, to perform their accounting. Without the discovery of logarithms, the slide rule would not have been invented (circa 1620) by Edmound Gunter (England). With the mathematics of logarithms, comprehensive tables were also published so that complex computations involving powers, roots and difficult division and multiplication became much easier and far less time consuming.

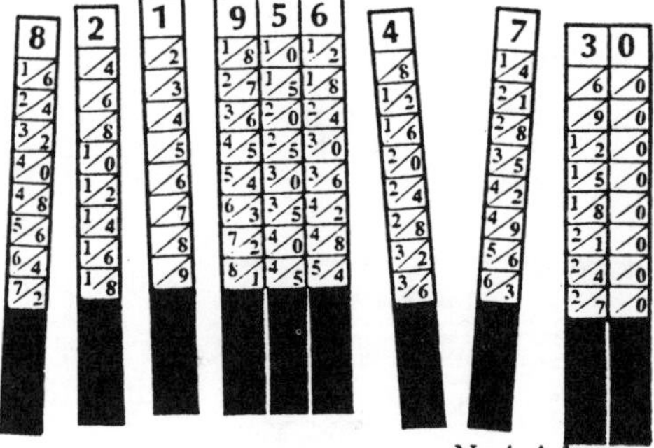

Napier's bones

In 1642, at the age of eighteen, French mathematician Blaise Pascal built the first calculator, probably to aid him in the accounting he was doing at his father's office. It could add and subtract, but it did not become popular from a commercial point of view since merchants could hire people to do the work for far less than the cost and maintenance of the machine. But it was an important step toward more sophisticated devices. In 1673, German mathematician Gottfried Wilhelm von Leibnitz was able to devise a calculator that could also do multiplication and division. These machines were far from perfect, but they were a very important beginning. They were continually improved and expanded, and ultimately evolved to the desk top calculator.

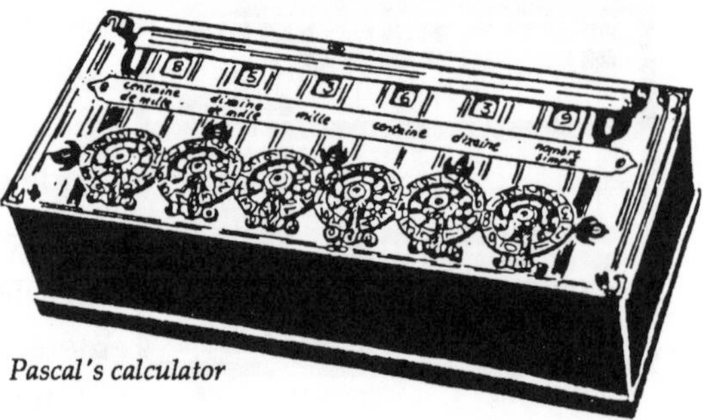

Pascal's calculator

In the meantime, Englishman Charles Babbage became very disenchanted with mathematical errors continually recurring in published timetables and charts, and decided to begin building a machine that could be programmed to process data to preset specifications (circa 1812).

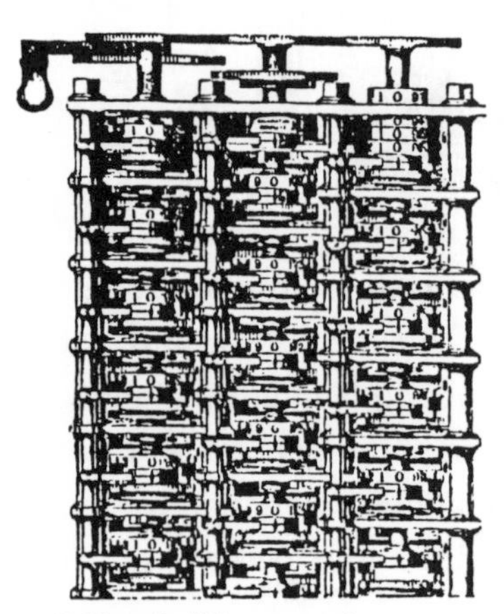

Babbage's difference engine

Unfortunately the technology of the time was not refined enough to produce the needed gears and cogs. Nevertheless, his work along with the computer programming work of Ada Lovelace[2] furnished the foundations for the modern computer.

The next major breakthrough came with population census taking. The 1880 census took 10 years to count manually. By the time it was finished it was about time for the next census. In 1887 the U.S. Census Department announced a contest open to the public to develop a reliable and efficient system for taking the census. Inventor Herman Hollerith entered a machine that processed information from punched cards using counter wheels and electromagnetic relays. He made the finals, but his machine met with much skepticism. Each of the three finalists was given a trial run to perform. Hollerith's machine did the job in 5.5 hours and the nearest runner up's method took 44 hours. He won the contest and the job of the 1890 census, which his machine completed in only one month! Even so, this machine was not capable of storing data and acting on that stored data, as envisioned by Charles Babbage.

The 20th century has been the dawn of the modern computer. Within the decades of this century and with the improvement of technology — the use of electricity to power machines, the development and use initially of vacuum tubes and later the use and discovery of the transistor and the integrated circuit, the development of the LSI (large-scale integrated) circuit on a silicon chip which made the personal computer feasible in size and price — computing devices have advanced enormously.

---

[1]Of course navigation required special types of calculating devices for charting courses by the stars. Here we find the ancient astrolabe and much later the sextant , 1757.

[2]As a tribute to the genius and work of Charles Babbage and Ada Lovelace, IBM built a working model of his Analytical Engine on which he and Lovelace (who furnished technical programming help and monetary support to Babbage) worked.

topological puzzle—*scissors, button & knot puzzle*

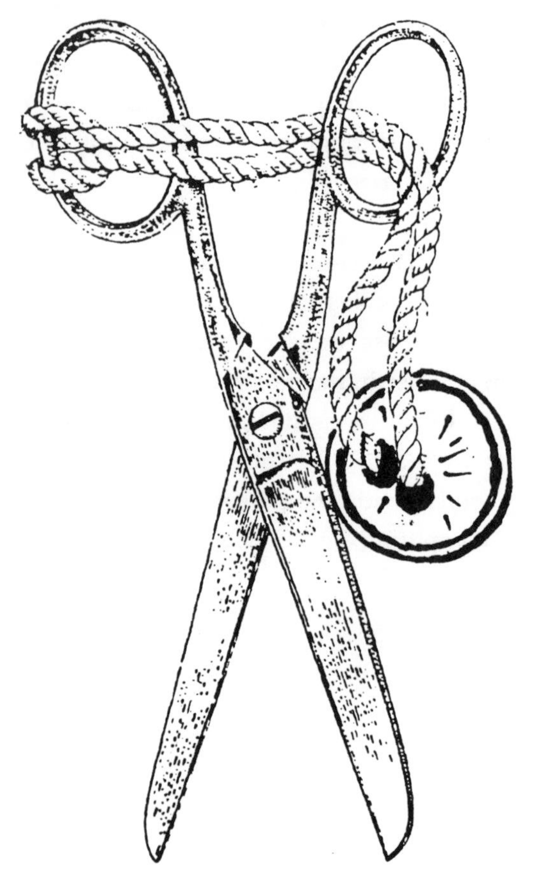

Take a pair of scissors, a button larger than the opening of the scissors' handle holes, and a cord. Thread through the button and scissors as illustrated.

Find a way to remove the cord without cutting or untying the cord.

a twist to the tower of Hanoi

This puzzle is adapted from the well known puzzle *The Tower of Hanoi.*

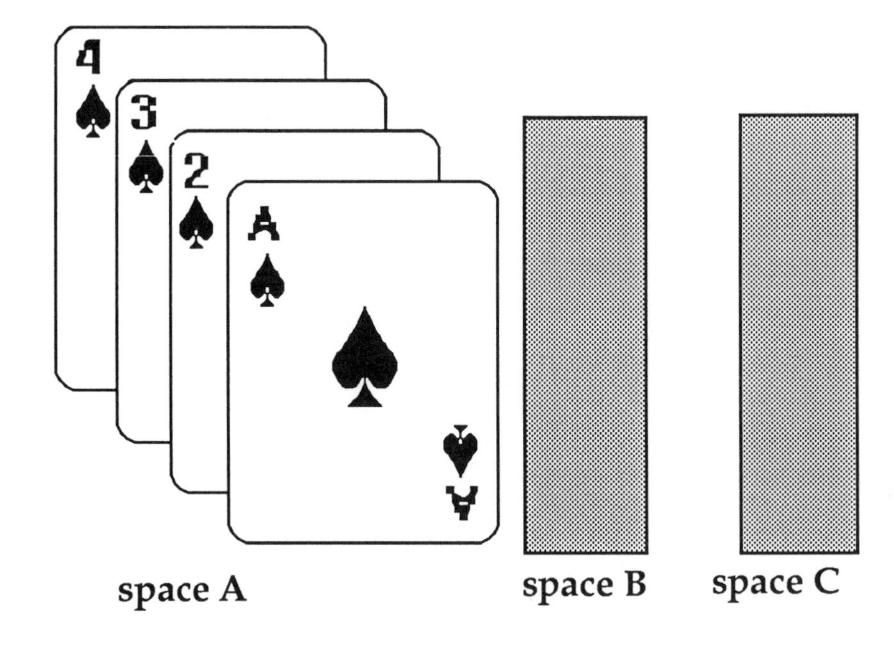

- Take four cards—an ace, a two, a three, and a four.
- The object of this puzzle is to relocate the cards in space A to space C conforming to the following rules:

  –*A larger valued card may never be placed on top of a smaller valued card. For example, you cannot put the two on top of the ace, but you can put the ace on top of the two, three or four.*

  –*You can only move a single card at a time to a new space.*

If you've mastered it with the cards ace through four, now add the five, then six ... and try it. Good luck!

impossible figures

Deceiving diagrams can often be stumbling blocks in solving mathematical problems.

For example, make a triangle with sides 17, 6.75 and 10.2.

A mathematical inspection of the sides of this "triangle", reveals the diagram is impossible using the given lengths. *(Any two sides of a triangle must be greater than the third side.)*

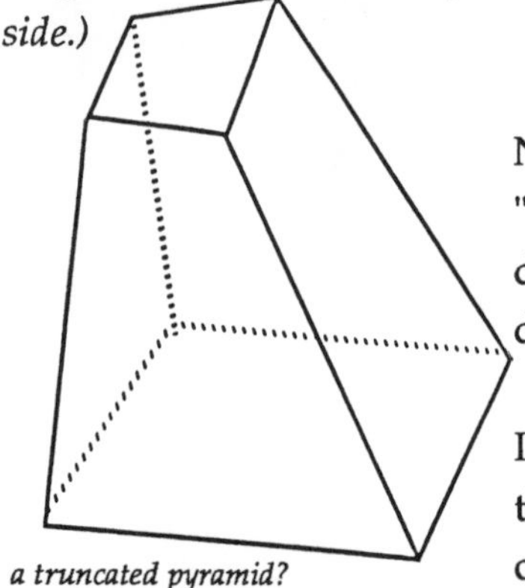

a truncated pyramid?

Now study the diagram of a "truncated pyramid". Can you discover why the diagram as drawn is not possible?

Impossible figures have stimulated the minds of artists and mathematicians for centuries. Perhaps the first impossible figures resulted from an artist's misinterpretation of perspective — or perhaps it was intentional. We find examples of such works as early as the fifteenth century in the restored work of the Grote Kerk in Breda in Holland. Below we see the use of three pillars and two arches —the pillars should be collinear, but the manner in which they are placed in perspective makes the arches bend forward and the middle pillar appear to be in the background. In the sixteenth century, Giovanni Battista Piranesi produced litho-

A representation of the three pillars in Kerk's work.

graphs —*Carceri d'invenzione* (imaginary dungeons) — in which impossible figures appear to create bizarre spatial scenes.

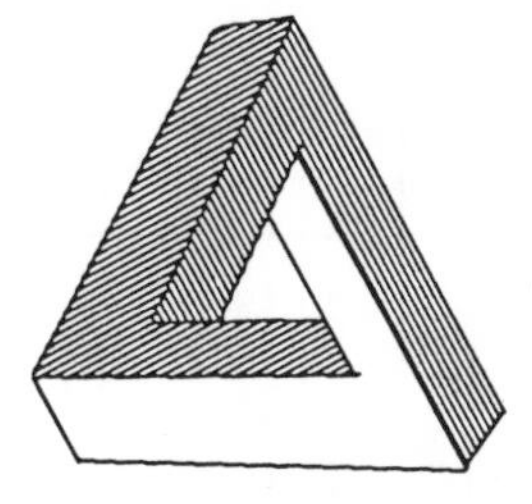

the tribar

In the nineteenth century we note the intense study and creation of optical illusions. And in the twentieth century we find exciting works involving impossible figures. The Swedish artist Oscar Reutersvård in the 1930's drew the first impossible tribar using an arrangement of nine cubes. He subsequently experimented and produced many drawings. Roger and L.S. Penrose in the 1950's wrote on impossible figures in which they described the tribar and the concept of an endless staircase which could be ascended and descended endlessly and yet remain on the same level. These ideas were embellished by such artists as M.C. Ecsher.

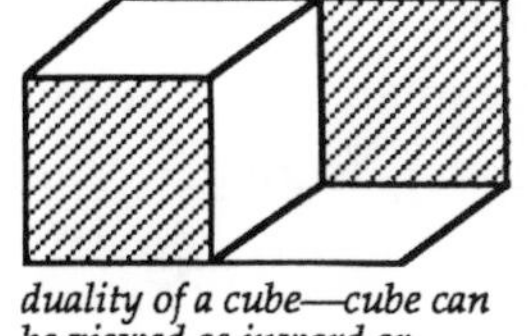

duality of a cube—cube can be viewed as inward or outward

Escher created many fascinating lithographs—*Waterfall* (1961), based on the impossible tribar—*Concave and Convex* (1955), using the convexity and/or concavity of the cube — *Belvedere* (1958), using the impossible cuboid — *Ascending and Descending* (1960), based on the staircase as presented by Roger and L.S. Penrose. These works have come to be studied, documented, analyzed and enhanced with special commentaries in books of author Bruno Ernst.

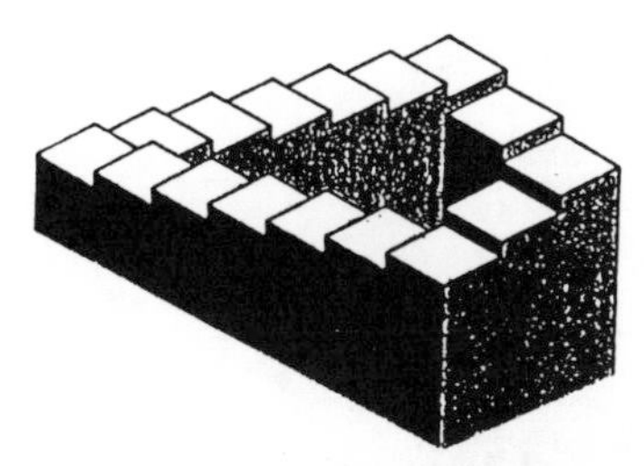

ascending/descending staircase based on Penrose's work

Today, computers are the new medium available to the artist/ mathematician. Computers, in the hands of the artist/ mathematician, are helping create the next generation of impossible figures. For example, look what happens to the impossible prong when, by the click of the mouse, it is reversed (make all white parts black and all back parts white). It now has a whole new feeling. It is exciting to anticipate these new figures. The fascination and mathematical explanations for the impossible figures of the future will definitely stimulate the mind and the imagination.

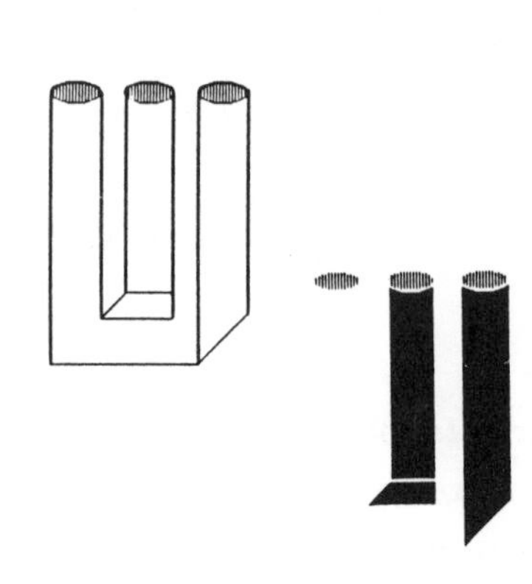

The impossible prong before and after being reversed using a computer.

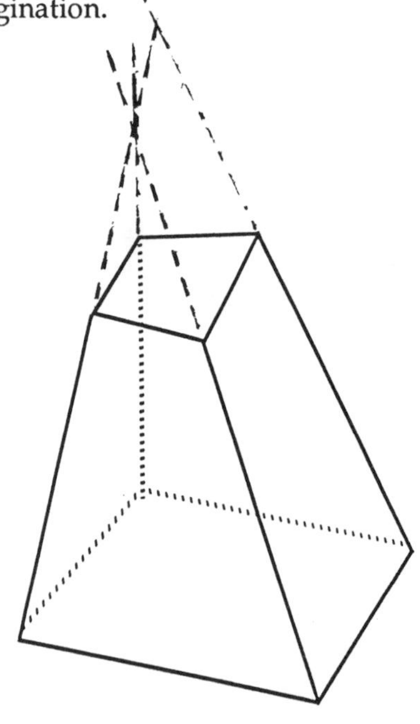

The extended lines do not meet in a point. This illustrates why the truncated pyramid on page 252 is impossible.

Which coin is counterfeit?

One of the twenty seven coins is counterfeit. The counterfeit coin weighs less than the other 26 coins, which are of equal weight. What is the least number of weighings needed to determine the counterfeit coin?

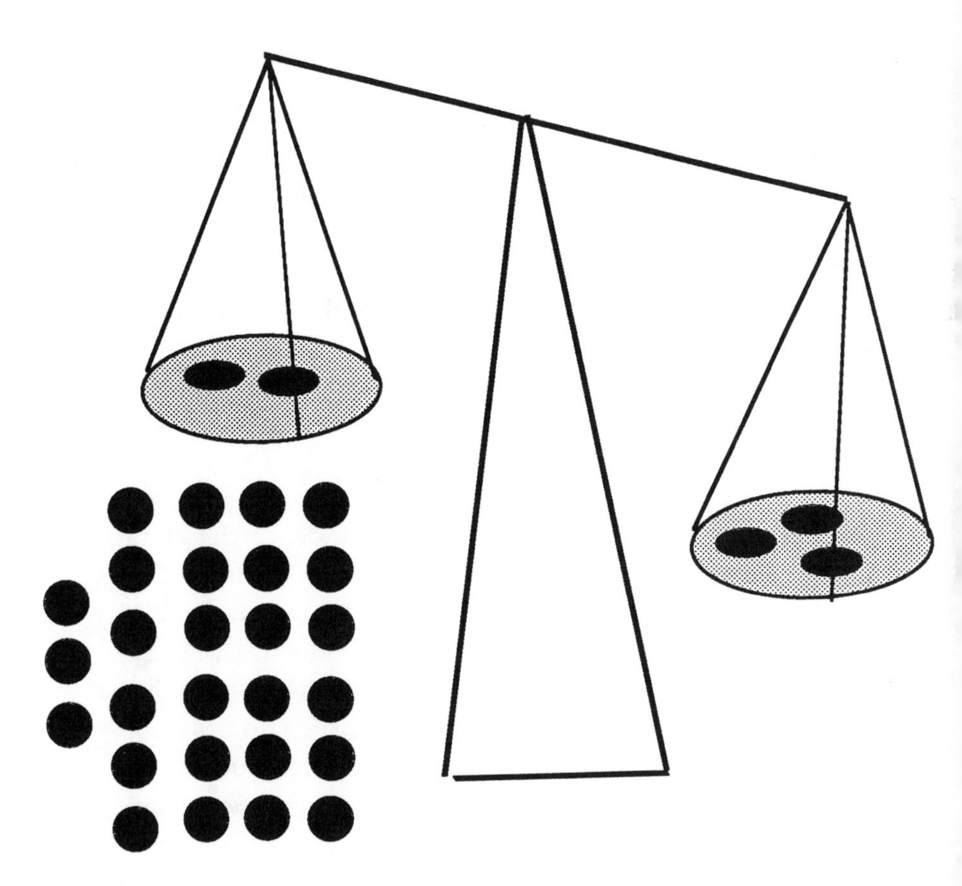

For solution, see the appendix.

mathematics, Moslem art & Escher

The Islamic religion played a predominant role in the evolution of Moslem art. Were it not for its strict edicts prohibiting the artist from using images of living things, their art would probably have had a totally different character. Moslems believed that Allah (God) was the one and only creator of life; and therefore an artist's attempt to paint or sculpt a life-like object would trespass into Allah's domain. This belief imposed unusually tight restrictions on the subject matter an artist of Moslem religion could use. Artists had to avoid the portrayal of human or animal forms in their works. If they ever appeared, they had to be very stylized. Consequently, artists channeled their creativity into specific areas. Their works were confined to ornamentation and mosaic, and concentrated on geometric designs and floral motifs (the arabesques). As a result Moslem artists delved into mathematics in order to broaden the realm of the medium to which they were restricted.

The Alhambra is an exquisite illustration of Moslem architectural and artistic developments. A palace and fortress located in Granada, Spain, The Alhambra was built by the Moors between 1248 and 1354, and is one of the finest examples of Moorish art in Europe. The walls of The Alhambra are decorated and emblished with an incredible variety of patterns. Examination reveals mathematical concepts as symmetries, tessellations, reflections, rotations, translations of geometric forms, congruences between dark and light patterns. The artists, in their quest to expand their art form discovered and used these mathematical concepts. These mathematicians/artists learned how to tessellate (cover) planes and discovered all the possible symmetries that exist in a plane. It was at The Alhambra that M.C.Escher was inspired to pursue his work with tessellations in art. After his first short visit in

1926, he labored but failed to design a tessellation with which he was satisfied. He put aside his efforts in space-filling art for ten years. In 1936 he returned to The Alhambra with his wife. On this visit he was again inspired and impressed by the wealth of spacefilling designs. He and his wife spent days copying these fascinating tessellations. When he returned home he studied them voraciously along with any other books he could find on this type of ornamentation and its mathematical foundations. He became so steeped in this subject that he formulated a complete system for himself by which he no longer had to struggle when creating a design – the laws of periodic space-filling did the work for him. In the words of Escher,

One of the many rooms of The Alhambra decorated with tessellations.

> *"The Moors were masters in the filling of a surface with congruent figures and left no gaps over. In the Alhambra, in Spain, especially they decorated the walls by placing congruent multicolored pieces of majolica together without interstices. What a pity it was that Islam forbade the making of "images". In their tessellations they restricted themselves to figures with abstract geometrical shapes....I find this restriction all the more unacceptable because it is the recognizability of the components of my own patterns that is the reason for my never ceasing interest in this domain."*

Yet it remains to be asked, if the Moslem artist had not been restricted, would their art, and especially the mathematics supporting it, have evolved ?

alquerque

Board games appear in almost all cultures. Developed as entertainment and pastimes, many games date back thousands of years, and are still enjoyed today. **Alquerque** is an ancient board game that has traveled between many countries over many centuries. Its origin dates back to ancient Egypt. In fact an engraved diagram of it appears on the temple of Kurna (circa 1400 B.C.). The Moors, who ruled most of Spain for five hundred years, first introduced Alquerque into Spain, hence the Spanish name for this checker-like game. Spanish settlers of Mexico brought the game with them, and thus a form of it is played by the Zuni Indians of New Mexico. Alquerque is a challenging game, requiring much thought, strategy and logic.

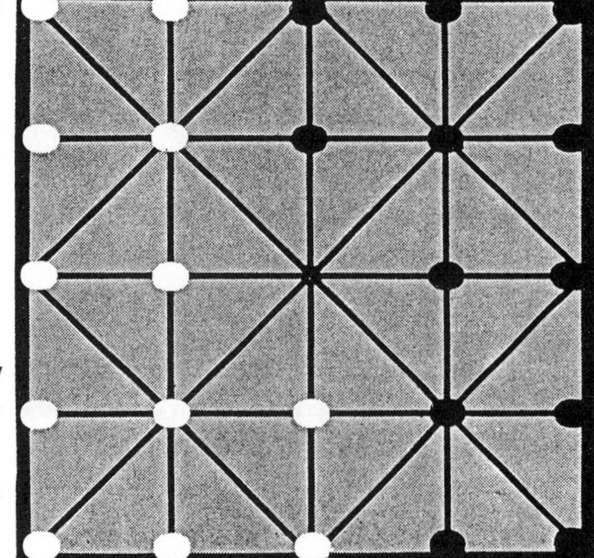

the game

1) The game is played by two players. Begin by arranging the 24 pieces as illustrated in the diagram.

2) Alternate turns.

3) A piece can be moved to any unoccupied adjacent space.

4) Pieces are captured and permanently removed by jumping an opponent's adjacent piece, and landing in an empty space adjacent to the piece jumped. A series of jumps in any direction is allowed.

5) If a player neglects to make a jump whenever possible, the opponent captures the player's piece.

6) The winner is the first player to capture all the opponent's pieces.

soroban—Japanese abacus

The Japanese abacus, the soroban, seems to be experiencing a renaissance in Japan. Soroban schools are flourishing, and some Japanese calculator manufacturers are producing both calculators and the soroban—sometimes mounted together.

The soroban advocates claim it leaves less room for error, is fast, if not faster, in the hand of an abacus artist than a hand held calculator; and it helps the user gain an understanding of arithmetic.

Annually, there are national soroban competitions. The competitors solve 20 problems, each involving the addition of twenty 11-digit numbers within five minutes.

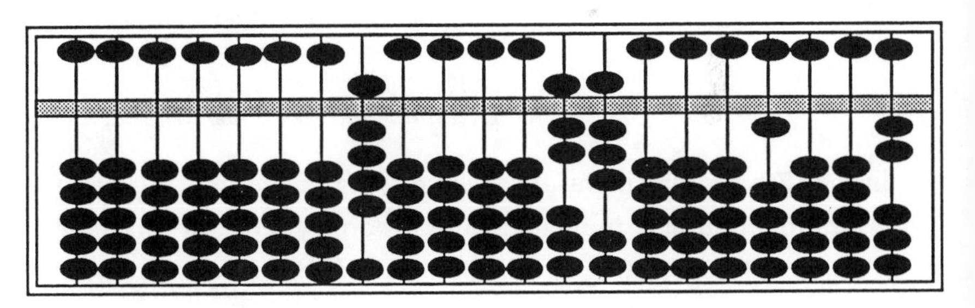

soroban—the Japanese abacus

Are curves & π always linked?

Ancient peoples from around the world knew about the special relationship between a circle's circumference and its diameter. π, the number representing the result when a circle's circumference is divided by its diameter, had been known and used on problems for thousands of years. Consequently, in order to calculate the area of a circle, π is also used, area=πr^2. Do the measurements of the lengths and the areas of all curved shapes involve π? In 450 B.C. Hippocrates studied lunes and proved that the area of the 2 shaded lunes equal the area of the shaded triangle.

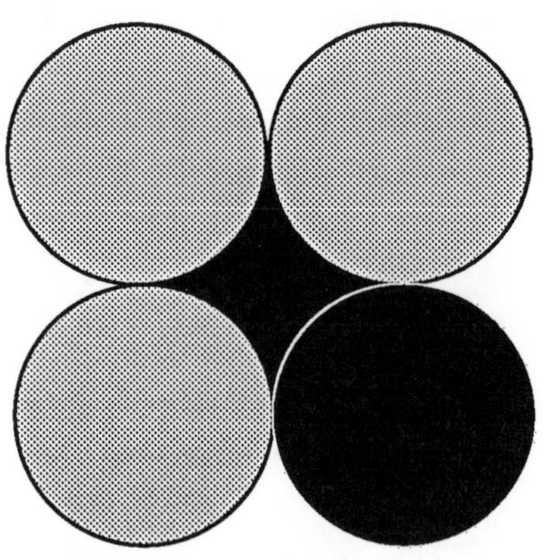

The urn shaped puzzle
Will the area of the shaded urn depend on π? The radius of each of the four congruent circles is 1/2 foot. Determine the area of the urn shape.

In the 17th century it was discovered that the length of a cycloid is a rational number independent of π, namely it is four times the diameter of the rotating circle. On the other hand the area under the cycloid arch is three times the area of the rotating circle, and thus involves π. These few examples illustrate that curves and π do not necessarily go hand in hand.

For solution, see the appendix.

> ***a treasure trove of geometric figures.***

One might view this diagram as a pleasant design made from lines forming different shapes. But a closer look reveals that the merging of two equilateral figures — a square and an equilateral triangle — produces this design. Many objects are hidden. In fact, it is only formed from one square and four equilateral triangles (each has a vertex at a corner of the square). Look for symmetries, rotations, squares, other equilateral triangles, pentagons, trapezoids, hexagons, right isosceles triangles, right triangles, rhombi, parallelograms.

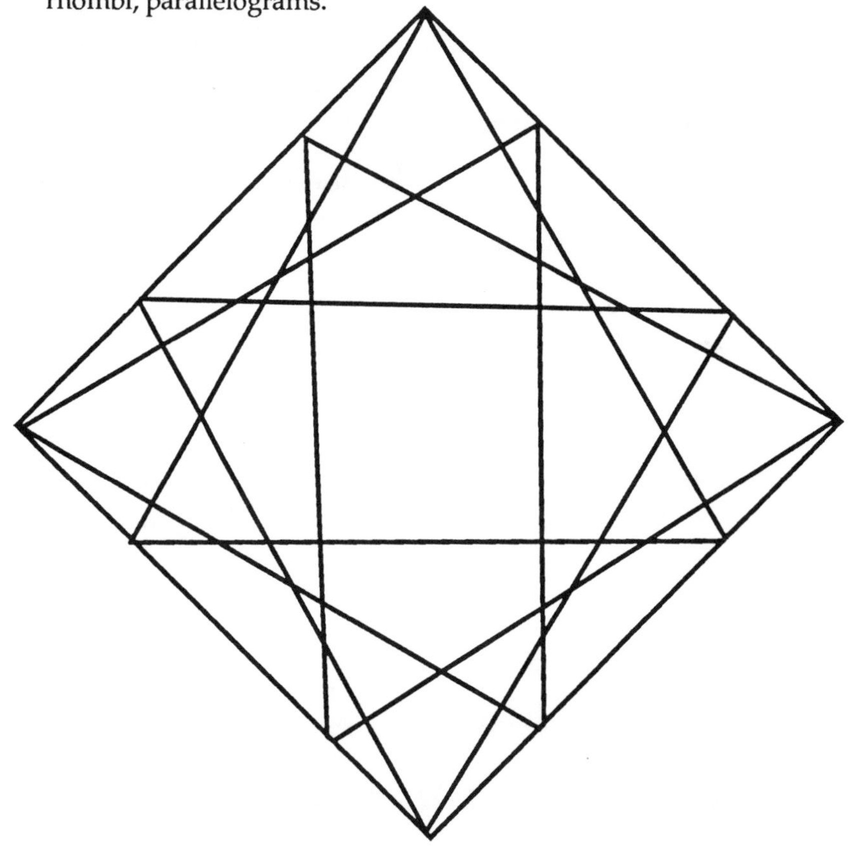

the game of reversi

The game *reversi* dates back to the 1800s, at which time it enjoyed its greatest popularity. Today the game is available commercially in various forms. One design uses cubes with the cube's six faces painted in different colors, thereby allowing up to six people to play. The game is also available in computer form with various levels of difficulty and speed in which you and your computer are at odds.

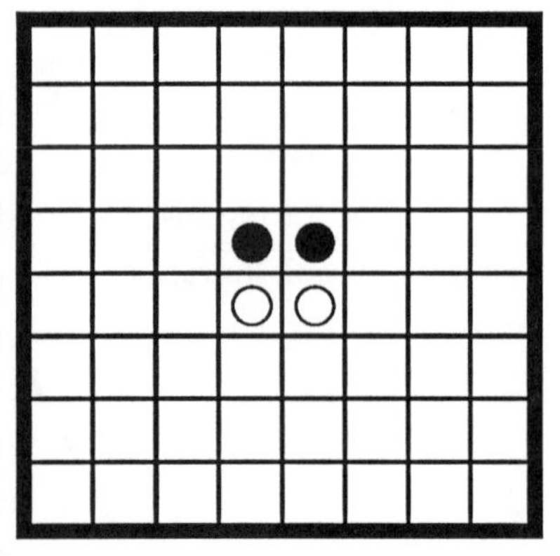

a possible start-up play for reversi

Reversi is played on a standard chessboard using 64 flat pieces with contrasting colors on each of the sides of the discs. Just as its name implies, one of the main movements of the game is reversing or flipping pieces. While the rules and procedure are simple, it can be a very challenging, fast moving game full of surprises until the end.

Procedure:

1) Each player begins with 32 pieces. The players select their color, and will alternate moves.

2) The first four moves must be played in the center four squares. Some of the possible start up plays are—

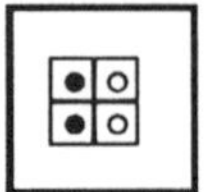

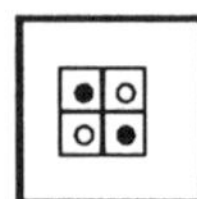

3) Each player alternately places his or her piece in an empty square adjacent to one occupied by an opponent in an attempt to "capture " the opponent's pieces.

4) A "capture" is made when an unbroken line of one or more pieces of an opponent is flanked on each end of the line. At this point the opponent's pieces are reversed and become the capturing player's color. (Note: a piece may change colors a number of times in the course of a game.)

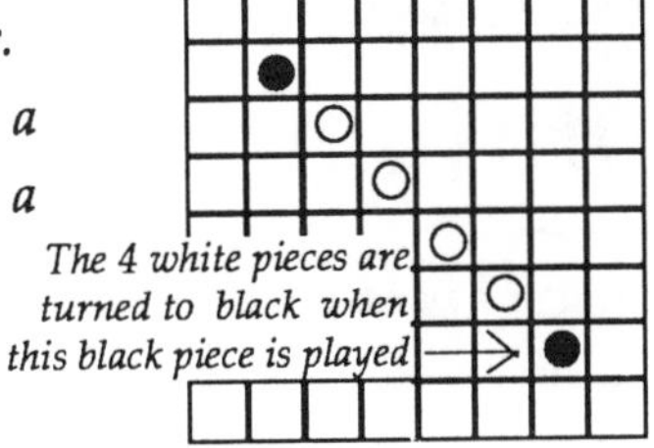

5) More than one line may end up being reversed when a single piece is played.

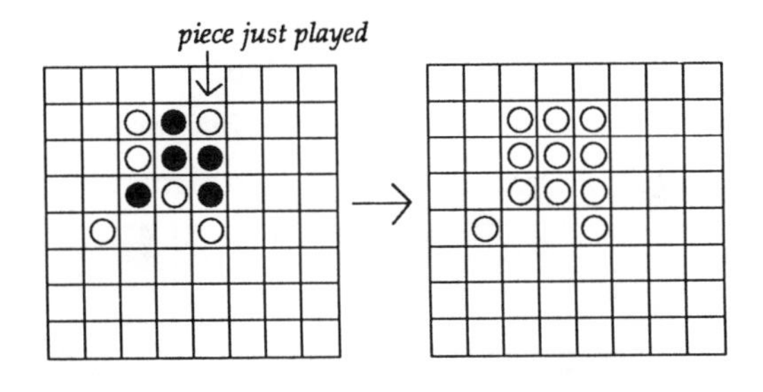

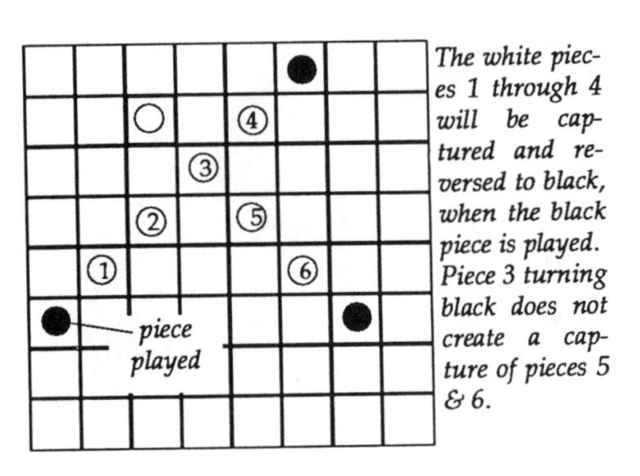

6) A player is not allowed to decide which pieces he or she would like to reverse. All must be reversed in the line or lines. Only the line(s) as a result of the piece played may be reversed, and no captured pieces may be used to create new captured lines in the same turn.

7) If a player cannot make a capture, he or she must miss that turn.

8) The game ends when all pieces have been played or no moves remain.

9) The winner is the player with the most colored pieces on the board.

ENJOY!

poet & mathematician –*Omar Khayyam*

Omar Khayyam (circa 1050?-1123?) is known to most people for his poetic and literary works, such as *The Rubaiyat*. But in the world of mathematics, he is known for many contributions including:

- the discovery of a geometric solution for solving cubic equations;
- a rule for determining the 4th, 5th, 6th and higher powers of a binomial (as mentioned in his book Algebra);
- writing a critical treatment of Euclid's Elements.

He was also an accomplished astronomer, and his reform of the Persian calendar made it almost as accurate as the Gregorian. It is in *The Rubaiyat* we find the passage which refers to his calendar reform.

Ah, but my Computations, People say,
Reduced the Year to better reckoning? – Nay,
'Twas only striking from the Calendar
Unborn Tomorrow, and dead Yesterday.

Leonardo da Vinci & the ellipse

Leonardo da Vinci was the ultimate Renaissance man — artist, architect, scientist, inventor, mathematician, philosopher, sculptor. His notebooks are full of sketches and innovative ideas spanning a variety of subjects, sometimes too advanced for his time. He invented various types of special compasses capable of producing parabolas, ellipses and proportional figures. He is also credited with the invention of the perspectograph, used by artists, such as Albrecht Dürer, to help draw objects in perspective. He strove to learn and master subjects whose understanding he knew was essential or influenced his work. His notes and various innovations have been used by artists to enhance and facilitate their work.

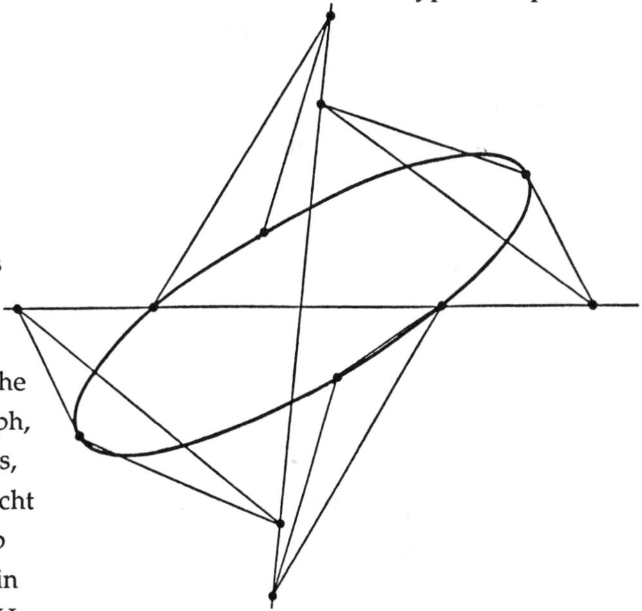

This ingenious way of tracing an ellipse was devised by Leonardo da Vinci.

Draw two intersecting lines.
Cut out a triangle. Assign one of the triangle's vertices to each line. Slide the triangle along the lines as illustrated. Mark the positions of the third vertex. The third vertex's marks will trace an ellipse.

The above sketch illustrates his creative and inventive way of tracing an ellipse.

ϕ—not one of your everyday irrational numbers

It's time for ϕ, phi, to be recognized in our daily use of numbers. Why is it rarely mentioned in everyday math conversation or computation? It definitely deserves a place of distinction and honor. ϕ is one of the mathematical concepts that appears in nature again and again —as do triple junction, c , e, i, π, tessellations, equiangular spirals, hexagons, platonic solids, helices, cycloids, fractals, symmetries. ϕ is not one of your everyday irrational numbers whose frequency of appearance in nature and in diverse mathematical ideas should rank it in popularity with π. Since it is inseparably linked to the golden rectangle and the Fibonacci sequence, it is automatically present wherever these mathematical concepts appear – growth patterns of flowers, pines cones, chambered nautillus shell.

Although it was not until the 20th century that the golden mean (also known as the golden ratio, golden section, golden proportion) received a symbol name, the discovery of ϕ, phi, dates back thousands of years. We know the ancient Greeks constructed ϕ and used the golden rectangle in their architectural designs (Parthenon) and in proportions of sculptures. Perhaps it was when the geometers of antiquity studied proportion and geometric means that they discovered how to construct the geometric mean of a line segment.

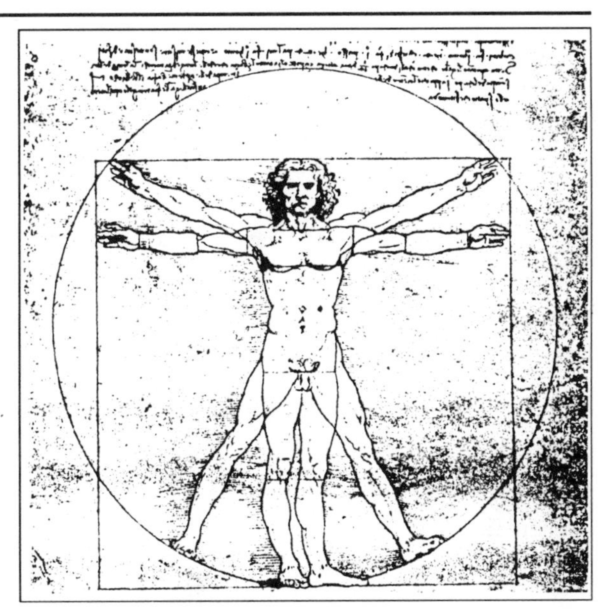

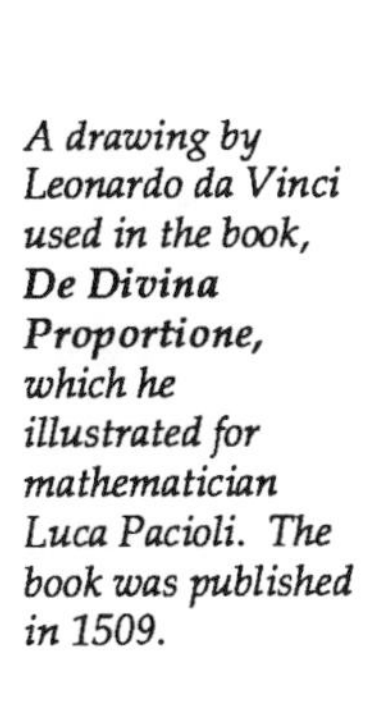

A drawing by Leonardo da Vinci used in the book, ***De Divina Proportione,*** *which he illustrated for mathematician Luca Pacioli. The book was published in 1509.*

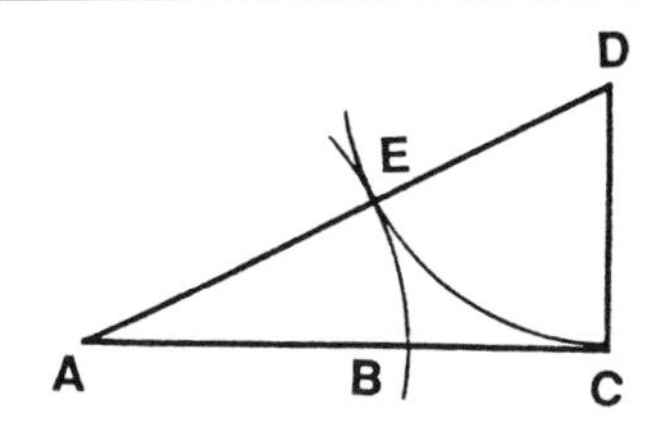

Dividing a line into the golden ratio

1) *Draw segment AC of any unit length, and call it one unit.*
2) *Construct segment CD perpendicular to AC and half its length.*
3) *Draw in segment AD.*
4) *Draw circle with center D and radius $|CD|$. It intersects AD at E.*
5) *Arc off length of segment AE on AC. Label this point B. B divides AC into the golden ratio so that $(|AC|/|AB|)=(|AB|/|BC|)=\phi\approx 1.618...$*

The discovery may then have led to the formation of the golden rectangle or vice versa.

Now that we have an actual value for $\phi=(1+\sqrt{5})/2\approx$ 1.618...., perhaps we will able be able to recognize it when it appears in such things as:

- *equiangular spiral*
- *pentagon*
- *golden rectangle*
- *golden triangle*
- *art*
- *architecture*
- *algebra*
- *infinite series*
- *Platonic solids*
- *inscribed regular decagon*
- *limit of sequence involving the Fibonacci numbers, ratio of consecutive terms of the Fibonacci sequence*
- *or whatever you may be dabbling in*

mathematics in the garden

"Good morning day!" exclaimed the gardener, as she greeted the sunrise and her plants. But little did she know that strange things were lurking in the leaves and rich soil. Deep in the roots of the plants were fractals and networks, and from the cosmos, irises, marigolds, and daisies Fibonacci numbers were staring at her.

She proceeded about her daily ritual of tending to her garden. At each place, something unusual appeared, but she was oblivious, captivated only by the obvious wonders that nature presented.

She first went to clear out her ferns. Removing the dead fronds to expose the new fiddle heads, she did not recognize the equiangular spirals greeting her and the fractal-like formation of leaves on the ferns. Suddenly, as the breeze shifted, she was struck by the lovely fragrance of the honeysuckle. Looking over, she saw how it was taking over the fence and getting into the peas. She decided it definitely needed some judicious pruning. She did not realize that helices were at work, and the left-handed helices of the honeysuckle had wound around some of the right-handed helices of the peas. It required a careful hand to avoid damaging her new crop of peas.

Next she moved to weed beneath the palm tree she had planted to give her garden a somewhat exotic accent. Its branches were moving in the breeze, and she had no idea that involute curves were brushing against her shoulders.

She looked over at her corn smugly "Ha!" she thought. She had been hesitant to plant corn, but was now encouraged by how well the young corn was progressing. Unbeknownst to her, triple junctions of corn kernels would form within the ears.

How well the entire garden was shaping up and exploding with new growth! Admiring the new green leaves on the maple tree, she knew there was something inherently pleasing in their shape — nature's lines of symmetry had done their work well. And nature's phyllotaxis was only evident to the trained eye in budding leaves on branches and stems of plants.

Glancing around, she focused on the carrot patch. She was proud of how they were doing, and noted they needed thinning to insure uniform good sized carrots. She did not want to rely on nature to tessellate space with carrots.

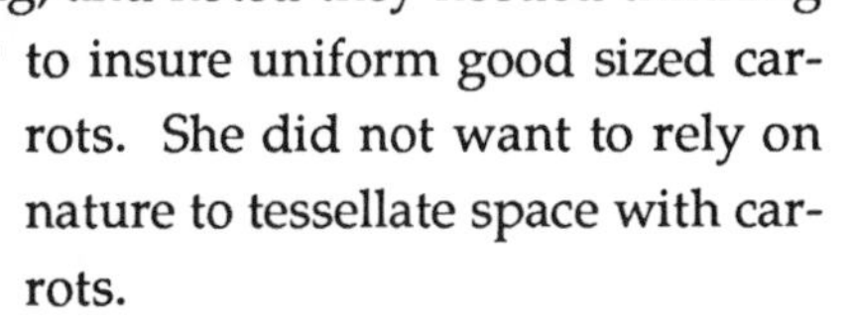

She had no idea that the garden abounded with equiangular spirals. They were in the seedheads of the daisies and various flowers. Many things that grow form this spiral because of how it retains its shape while its size increases.

It was getting warm, so she decided she would continue the cultivation when the sun shifted. Meanwhile, she made one final assessment — admiring the combination of flowers, vegetables and other plants she had so thoughtfully selected. But once more something escaped her. Her garden was full of spheres, cones, polyhedra and other geometric shapes, and she did not recognize them.

As nature puts forth its wonders in the garden, most people are oblivious to the massive calculations and mathematical work that have become so routine in nature. Nature knows well how to work with restrictions of material and space, and produce the most harmonious forms. And so, during each day of spring, the gardener will enter her domain with a gleam in her eye. She will seek out the new growth and blossoms each day brings, unaware of the mathematical beauties flowering in her yard.

puzzles to tax the mind

The Pier & Plank Problem: A detached fishing pier (its bridge was destroyed) is twenty feet from the main pier. Both are the same height. Some eager fisherpeople want to try their luck on the detached pier. They have two planks, each 19.5' long and 1' wide. They cannot nail them together. One clever person finds a way to use them for a crossing. How were the planks used to reach the detached pier?

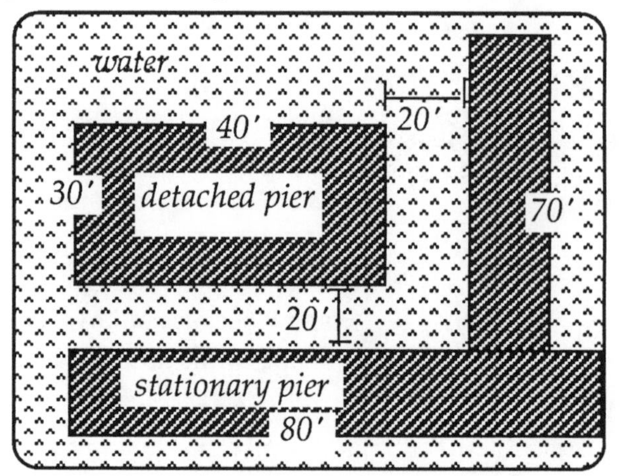

Logic tester: Martha, Bill and Terry each have two different occupations, which are also different from one another's. Their occupations are writer, architect, teacher, doctor, lawyer, and artist. Each character in each statement is a distinct person.

1. *The teacher and writer went skiing with Martha.*
2. *The doctor commissioned the artist to paint a mural.*
3. *The doctor had a meeting with the teacher.*
4. *The artist is related to the architect.*
5. *Terry beat Bill and the artist at chess.*
6. *Bill lives next door to the writer.*

Find each person's occupations?

For solutions, see the appendix.

Galileo's experiments pay off —cycloid discoveries

The 17th century was a time of interest in the mathematics of mechanics and motion. It was the time of the cycloid—*the curve traced by the path of a fixed point on a circle which rolls smoothly on a straight line.* Galileo (1564-1642) was one of the many prominent figures interested in the cycloid. He discovered, but did not prove, two

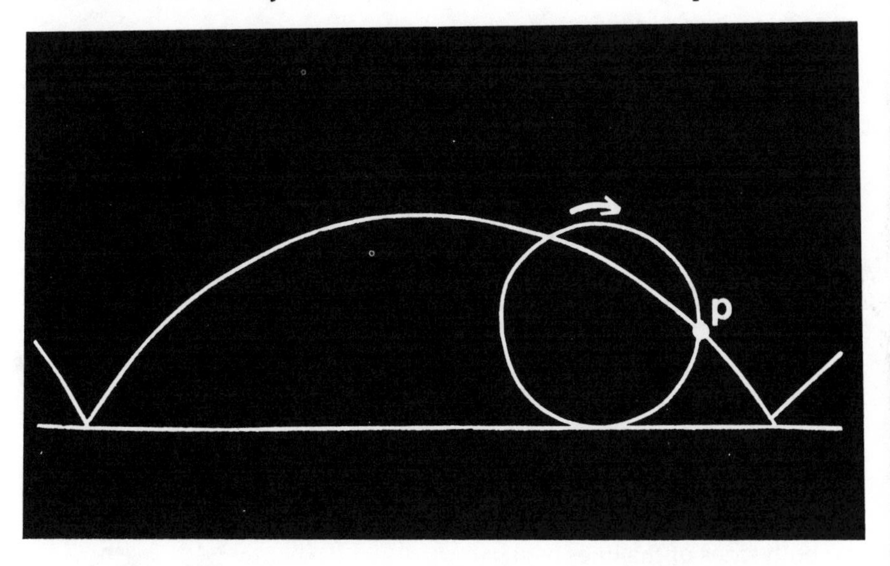

important facts about the cycloid. He found that the length of one arch of the cycloid is 4 four times the diameter of the revolving circle. He discovered this by physically measuring the arch traced with a cord and comparing this to the circle's diameter. Studying the area enclosed under the arch of the cycloid, he cut out the area with a sheet of lead and weighed it, compared it to that of the revolving circle, and determined the area was three times the area of the circle. His experiments proved to be exact. Unfortunately, the mathematics of the time prevented proof of these discoveries.

mathematics & design

Designs—be they in Native American rugs, Japanese mons, stone carvings from prehistoric times, pottery, tapestry, intricate works on Moslem buildings, or computer generated—all illustrate a wealth of mathematical ideas. In some cases the artists were unaware of the mathematics behind their designs, while others relied on mathematics to create them. Regardless, mathematical concepts are present and interesting to discover, explore, and identify.

For example, this rug design, fashioned from works of Native Americans, possesses—

lines of symmetry and reflections – Vertical and horizontal lines running through the middle divide the design into identical parts and identical designs which are reflected on both sides of the lines.

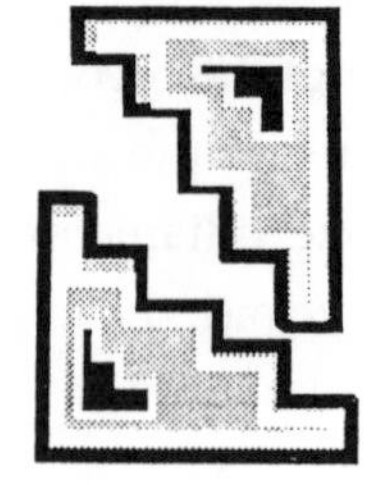

congruences – Congruent shapes are apparent throughout.

similarities and proportions – Geometric objects of the same shape but varying size are balanced through the design.

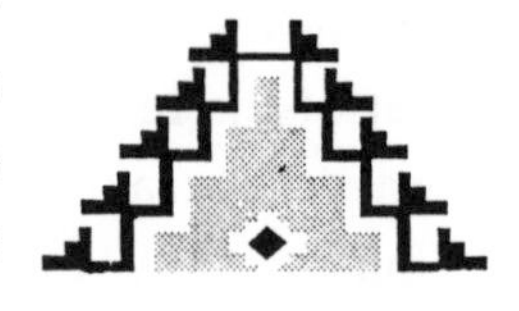

Japanese family crest designs (mons) became a tradition around the eleventh century. Many feature stylized plants, animals, natural treasures, mathematical objects and ideas. Mathematical motifs in mons include —*symmetry, knots, triangles, squares, circles, cubes, solids, hexagons* — all drawn in such a way that the impact is bold and dynamic.

According to Moslem teachings, artists trespassed into Allah's domain if their works took on a life-like appearance. Consequently, they developed and used mathematical concepts to enhance and expand their art form. Here *tessellations, rotations, transformations, translations* of geometric patterns abound, as well as symmetries and congruence. In more recent years M.C. Escher evolved and mastered the concepts of tessellations and adapted those ideas to living forms.

The *hypercube* and other 4th-dimensional designs were adapted by architect Claude Bragdon (circa 1913) in his works. An example is his Rochester Chamber of Commerce Building. In addition, he used *magic lines* in architectural ornaments and graphic designs. Magic lines are patterns formed from magic squares when the numbers of the square are sequentially connected, for example—

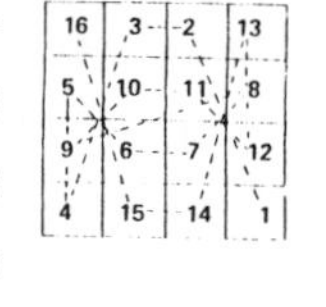

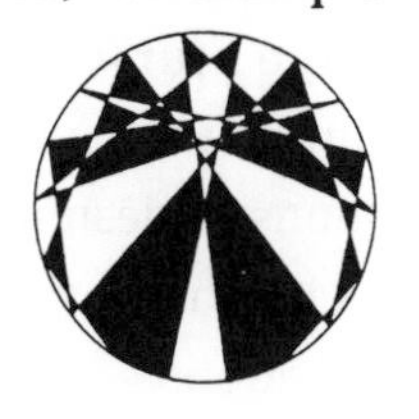

Modulo arithmetic has been used to produce some very bold graphic designs, such as this one—

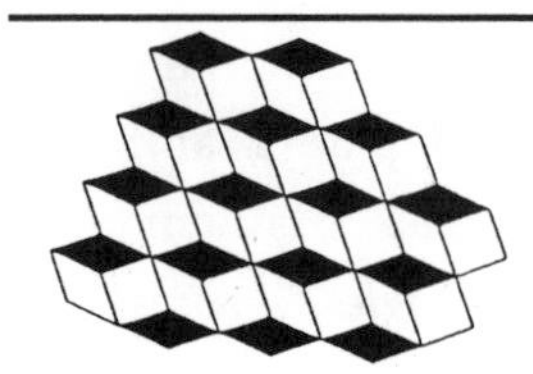

Optical illusions and *impossible figures* have been used in many forms of graphics. Here is one tantalizing design illusion.

Mazes, Jordan curves, and *knots* have been used in such designs as the ancient coin design from Knossos, Crete; the maze on this Navajo blanket, and spiral mazes in this stone carving from Ireland.

coin design from Knossos

stone carving from Ireland

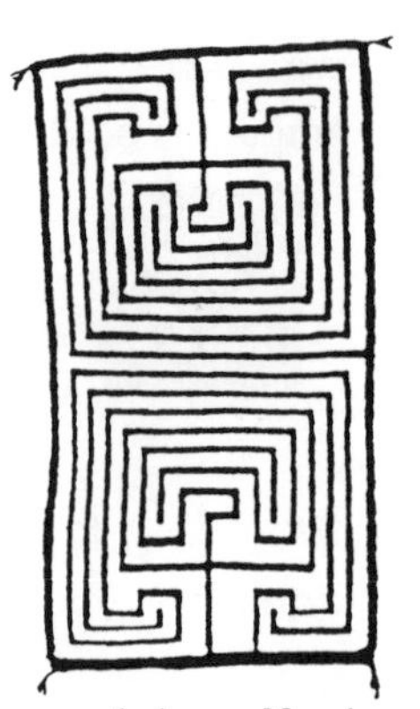

maze design on Navajo rug

Golden rectangle & ratio and equiangular spirals are other mathematical ideas that appear in designs such as the pentagram and various fabric designs.

Meanders appearing on pottery provide wonderful examples of *congruences, reflections,* and *mazes.*

Most recently the use of computers and the mathematical concept of *fractals* have been used to generate such exciting designs as tree-shaped fractals, dragon-curves, and snowflake curves.

Many artists relied on their understanding of mathematical ideas to create certain designs. Some of these artists include—Phidias, Leonardo da Vinci, Albrecht Dürer, George Seurat, M.C. Escher, Pietter Mondrian. Mathematics abounds with patterns and objects that lend themselves to the creations of designs. Graphic artists armed with mathematical concepts have a foundation from which to draw and enhance their creations.

Leonardo da Vinci's handwriting

Leonardo da Vinci was a genius of incredibly diverse curiosity and interests. His works, manuscripts and voluminous notebooks illustrate his talents as a painter, sculptor, architect, mathematician, scientist, engineer, and musician. Leonardo wrote and painted with his left hand. His signature and most of the entries in his notebooks run from right to left, and resemble mirror writing. When he was in his creative mode, he often shortened or left out words or phrases and wrote haphazardly to keep up with his thoughts. Occasionally in letters and work intended for others, he wrote from left to right. These appear awkward and the script does not seem to flow.

The numbers above illustrate his mirror writing.

Upon his death, the manuscripts in his possession were left to Francesco Melzi, his pupil and disciple. Melzi treasured the works and painstakingly catalogued them, but upon Melzi's death many of the works and unpublished notebooks were lost to thieves and dealers.

mathematics & the spider's web

The spider's web is a simple, but elegant natural creation whose aesthetic beauty can be mesmerizing when, dew laden, it glistens in the early morning sun. When mathematics attempts to describe this beautiful structure, the complexity of formulas needed is surprising.

There are many web designs that various species of spiders weave. They can be in the shape of sheets, triangles, funnels and domes. Looking at the webs of the Orb spiders reveals mathematical concepts which one would not easily guess could be linked in such a work of architecture.

The first mathematical objects one might notice are the two spiral-like curves of the web. Call the strands emanating from the center radii. The spiral-like curve is formed from chords connecting two consecutive radii. The chords placed between two consecutive radii appear parallel, and, assuming they are, all the corresponding angles along the radii would be congruent. If the spider's web had infinitely many radii, the chords would be single points rather than entire segments. Thus, instead of a jagged type of spiral, a smooth curve would be formed. Such a curve is a logarithmic spiral.

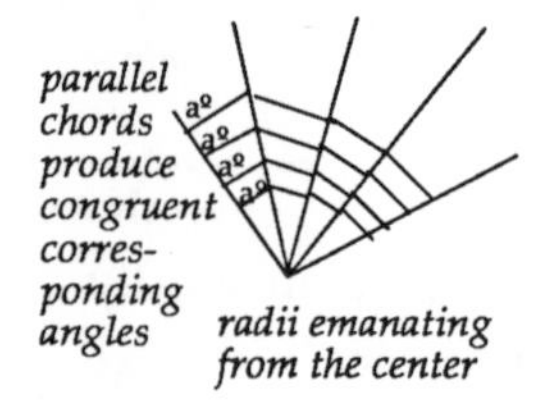

Properties of the logarithmic spiral

- Congruent angles are formed by tangents drawn at the points where the spiral intersects a radii. This is why the logarithmic spiral is also called an equiangular spiral

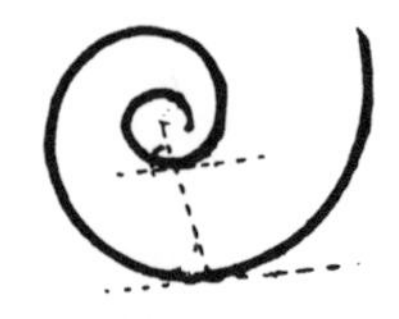

- The radii are cut by the spiral into sections that form a geometric progression. The spiral increases at a geometric rate, which is where it gets its logarithmic name.

- While the spiral winds its size alters but not its shape.

- If a thread is in the form of the spiral and its end from the pole (the center) is unwound, always keeping the thread taut, the endpoint of this thread will also trace a logarithmic spiral as it is unwound.

- The spiral-like web fills space economically, regularly, with strength flexibility and minimal materials.

How the spider constructs the web.

The spider initially sets up a triangular frame for its web. This minimizes the use of silk yet yields maximum strength and flexibility. The second spiral the spider forms is the main part of its snare. It is spun from the outer end toward the center, using the sticky silk. Both spirals it weaves are logarithmic.*

When the morning dew dramatizes the spider's web with its tiny droplets of clinging water (especially due to the sticky silk), the chords of the web bend under their load of droplets, thereby transforming each chord in to a catenary curve!

A *catenary curve* is a curve formed from a flexible chord or chain which is allowed to hang freely. Its general equation is $y=(a/2)(e^{x/a}+e^{-x/a})$, where a is the y-intercept.

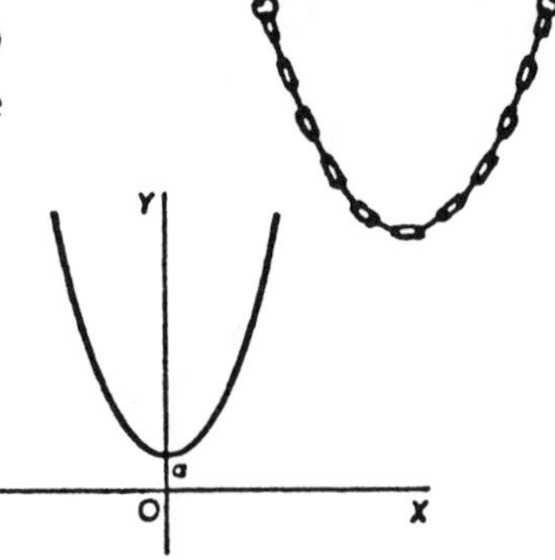

Since the irrational and transcendental number e—

$e = \lim(1+1/n)^n$ or

$1+(1/1) + (1/1\bullet2) +(1/1\bullet2\bullet3) + (1/1\bullet2\bullet3\bullet4) +\ldots=2.7182518\ldots$

— appears in the equation of the catenary curve, e is also a victim trapped in the spider's web. Thus, many and complex mathematical concepts —radii, chords, parallel segments, triangles, congruent corresponding angles, logarithmic spiral, catenary curve and e—are woven together in the spider's snare.

---

*The spider begins weaving the web using one of many silk producing glands. Some glands produce a sticky silk and others non-sticky silk. The frame, the radii and the first (the temporary) spiral are from non-sticky silk so the spider will not trap itself. The spider remembers all facets of its web, so that when it traps its prey it can immediately estimate its size and determine its location by sensing the vibrations of radii it plucks. The spiders eyes are not functional for viewing its prey. It quickly makes its way to its victim via the non-sticky radii, and finishes trapping its victim.

a sophisticated connect-a-dot

Here is a sophisticated connect-a-dot. The object is to connect consecutive prime numbers written in different number systems to reveal the hidden figure even while you test your knowledge of various number systems. In case you need to brush-up on your skills, the systems represented are Maya, Roman, Chinese script, Chinese rod numerals, Babylonian, Egyptian hieroglyphic, Egyptian hieratic base two, Greek, 15th century Hebrew, Hindu-Arabic numerals, and base two.

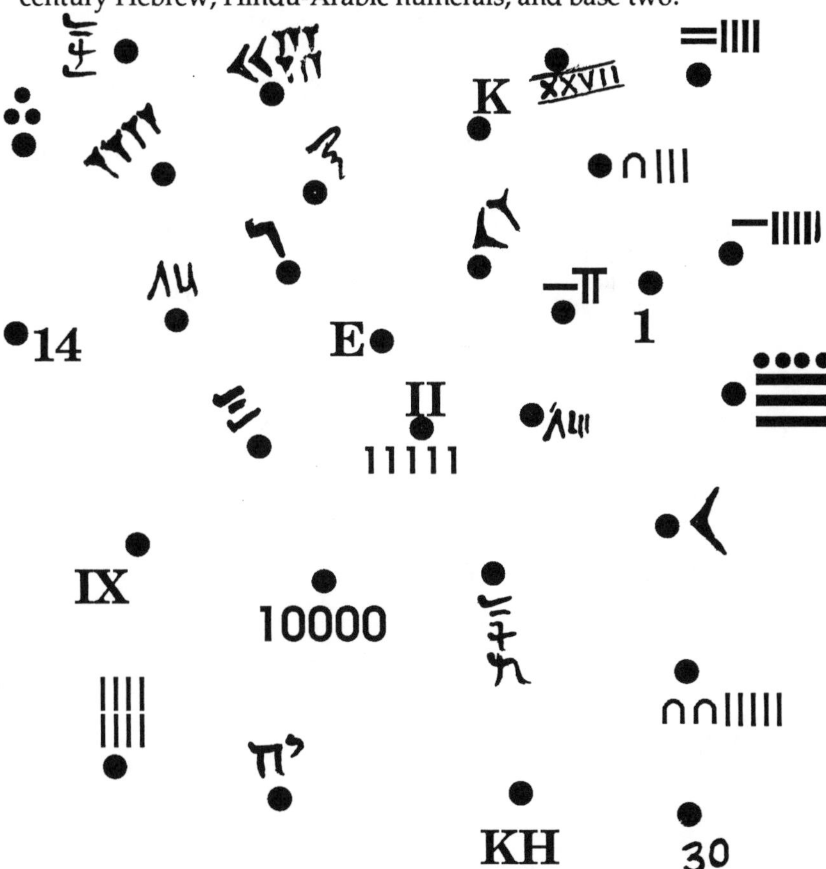

For the solution, see the appendix.

APPENDIX—solutions • answers • explanations

page 13—

bookworm puzzle: The answer is 10.5 if you understood the problem to read that the worm ate both the front cover of volume 1 and the back cover of volume 6 and all of volumes 2, 3, 4, and 5.

the bamboo pile: One solution is

The solution was submitted by Len Beyea.

musical coins puzzle: 5 moves— one solution is 6 to 3. 1 to 6. 5 to 1. 2 to 5. 3 to 2.

page 29—*solitaire checkers puzzle:* Minimum number of moves 15. Can you do better?

page 66—

the 8 checkers puzzle.

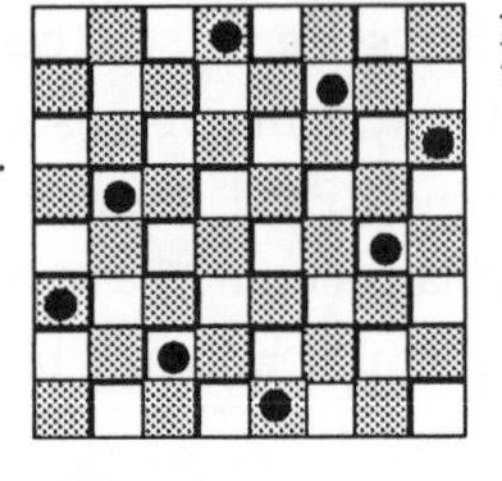

page 70— *Lewis Carroll's window puzzle*

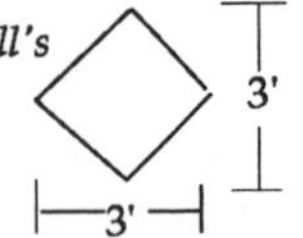

page 99—*puzzling scales:* One way to solve the puzzling scales problem is to use algebraic equations.

Let: b=weight of a bottle; g=weight of a glass; p=weight of a pitcher; s=weight of a saucer. Each weighing now can be represented by an equation. (1) b+g=p; (2) b=g+s; (3) 2p=3s. We want to find b=?g. Transform equation (3) to (2/3)p=s. Replace (2/3)p for s in equation (2), and we get b=g+(2/3)p. Call this equation (4).
Simplify this equation and solve for b in terms of g, and we get b=5g. Therefore, 5 glasses balance 1 bottle.

page111—*Every triangle is isosceles?* ΔABC is not every triangle. The diagram for step 3 makes a false assumption. It assumes that for any triangle the intersection of ray BD & line EF is in the interior of the triangle. This is not the case. For example,

B
D F
E

page 149—*creating triangles puzzle:*

page 107—*making rectangles:* 20

page 128— *measurement problem:* For distances from 1 to 40 units you could use straightedges of lengths 1, 3, 9, and 27.

page 131—*Romeo & Juliet puzzle:* The minimum number of turns that Romeo has to make to reach Juliet, after visiting every square only one time, is 15 including the first turn he makes when starting. Here is his path—

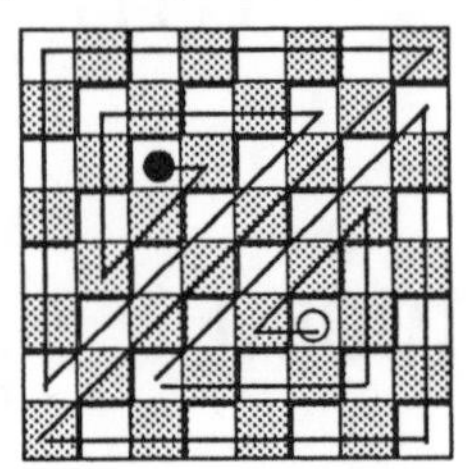

page 167— *overlaping squares problem;*

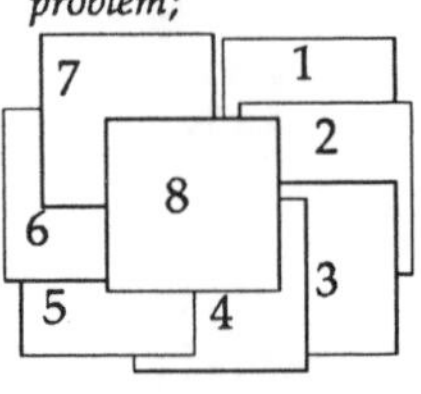

page 181— *shapes & colors puzzle:*

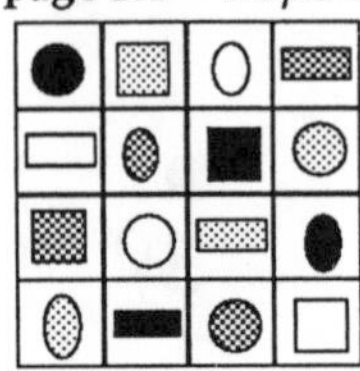

page 184—*the dock problem:*
No, it is not possible. Its network has four odd vertices.

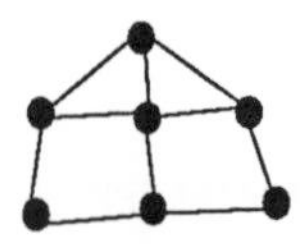

APPENDIX—solutions • answers • explanations

page 186—*the water jug problem:* First fill up the 8 liter jug and fill the 5 liter from it. From the 5 liter fill up the 3 liter jug. Pour the cider from the 3 liter into the 8 liter jug. Pour the cider from the 5 liter into the 3 liter. This now leaves 6 liters in the 8 liter jug, 2 liters in the 3 liter jug and the 5 liter jug is empty. Refill the 5 liter using the 8 liter jug. Fill up the 3 liter jug using one liter from the 5 liter. This now leaves 4 liters in the 5 liter jug. Pour the 3 liters from the 3 liter jug into the 8 liter jug which just had 1 liter in it. Now we have 4 liters in the 8 liter jug, and each party has its 4 liters.

page 195—*encircling the Earth:* Assuming the Earth's circumference is 25,000 miles, this converts to 132,000,000 feet. Adding an additional yard, the circumference of the rope in feet is 132,000,003 feet. Using the formula for the circumference of a circle, C=diameter • π. The radius of the Earth comes out to be about 21,008,452.49 feet. The radius of the rope, which is encircles the Earth, comes out to be about 21,008,452.97 feet. Their difference is .48 feet which coverts to 5.76 inches.

page 201—*puzzles to exercise the mind:*

dissection puzzles:

the stick puzzle:

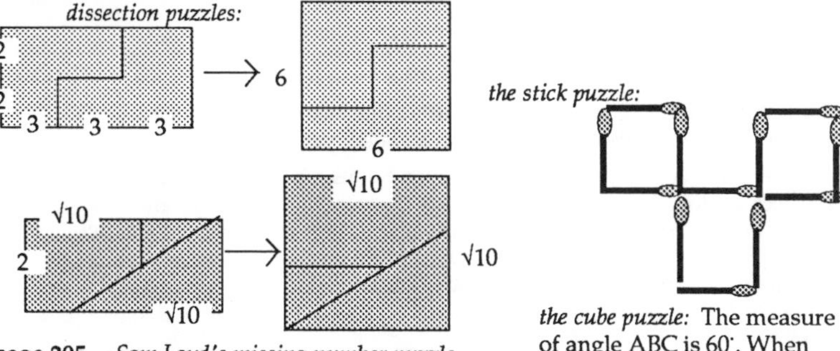

the cube puzzle: The measure of angle ABC is 60˚. When diagonal AC is added to the diagram, we see that ΔABC must be equilateral, since all the diagonals of the faces of the cube are congruent.

page 205— *Sam Loyd's missing number puzzle*

```
         853
749 | 638897
      5992
       3969
       3745
        2247
        2247
```

page 212—*the 7, 11, & 13 number oddity:* When abcabc is written in place value form, we get 100,000a+10,000b+1000c+100a+10b+c=100,100a+10,010b+1001c. Since 7x11x13=1001, it is a factor of 100100, 10010 and 1001. Therefore, it will always be divisible into abcabc. Any combination of products of 7, 11 and 13 will also be factors of abcabc. These are 77, 91, 143 and 1001.

page 217—*measuring old sayings with different units:* Give him an inch and he'll take a mile. — He demanded his pound of flesh. — A miss is as good as a mile. — The perfect 36". — Every inch a king. — 5 foot two, eyes of blue. — I love you a bushel and a peck. — An ounce of prevention is worth a pound of cure. — First down and 10 yards to go.

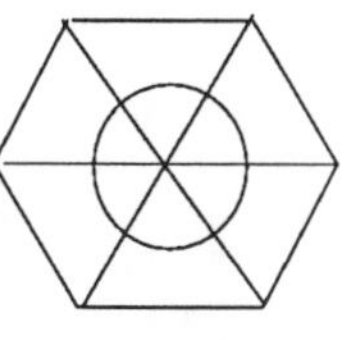

page 225—*tethered goat puzzle:* The area of the hexagon, composed of 6 congruent equilateral triangles is 12π acres. The area of the portion the goat can graze on is π acres (which is half the area of its fenced equilateral triangle). The circle's area is half the hexagon's area. Thus the area of the circle is 6π acres or 261360π square feet. The radius of this circle would be the length of the goat's rope. So, $261360\pi=\pi r^2$. $261360=r^2$.
$r\approx511$ feet.

APPENDIX—solutions • answers • explanations

page 255—*Which coin is counterfeit?*
The least number of weighings is 3. Divide the coins into three groups of nine coins. Use the first weighing to decide which group of nine has the counterfeit coin. Then divide this group into three groups of three coins. Use the second weighing to decide which group of three coins has the counterfeit coin. Finally, use the third weighing to decide which of the three coins in the group of three is counterfeit.

page 260—*the urn puzzle:*

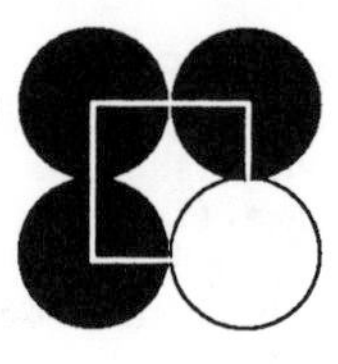

The area of the urn shape is 1 square foot. Since each side of the square is 1 foot, its area is 1 square foot. There are 4 quarter circles in the square, which make up one entire circle. The urn shape is made up of a circle and the portion of the square that is not covered by the four quarter circles. Therefore, its area =
=squares's area - the four quarter circles' area + the circle's area
=the square's area=1 square foot.

page 270— *puzzles to tax the mind:*

Logic tester: Statement 5 tells us that Martha must be an artist.

Using statements 1, 4, and 2, we determine that Martha is not a teacher, architect, writer, or doctor. That makes her a lawyer.

Statement 6 tells us that Bill is not a writer. So his jobs can be teacher, architect, or doctor. From statement 3 we deduce that the doctor and the teacher are different people. This means that Bill must be an architect. Bill cannot be a doctor, otherwise Terry would be a teacher and writer, which statement 1 indicates is not possible. We know Bill is not a writer from statement 6. Thus Bill's other job is a teacher, and Terry is both a doctor and a writer.

Martha—artist & lawyer.

Bill—architect & teacher.

Terry—doctor & writer.

the pier & plank problem:

pager 279— *sophisticated connect-a-dot:*
A five pointed star is formed.

1= 1 Hindu-Arabic
2= II Roman
3= 三 Chinese script
4= 𒐉 Babylonian
5= E Greek
6= ו Hebrew
7= [illegible] Egyptian hieratic
8= 𓐁 Egyptian hieroglyphic
9= IX Roman
10= 𒌋 Babylonian
11= 𒌋𒁹 Babylonian
12= ∧ч Egyptian hieratic
13= ∩IIII Egyptian hieroglyphic
14= 14 Hindu-Arabic
15= —IIIII Chinese rod numerals
16= 10000 base two
17= —Π Chinese rod numerals
18= יח Hebrew
19= [illegible] Mayan
20= K Greek
21= 二十一 Chinese script
22= [illegible] Mayan
23= ∧ч Egyptian hieratic
24= =IIII Chinese rod numerals
25= ∩∩IIIII Egyptian hieroglyphic
26= 𒌋𒌋𒐋 Babylonian
27= XXVII Roman
28= KH Greek
29= 二十九 Chinese script
30= 30 Hindu-Arabic
31= 11111 base two

BIBLIOGRAPHY

It would not be possible to form a complete bibliography for *More Joy of Mathematics* or for *The Joy of Mathematics*. Even listing all the books in my library would not suffice, for many sections were written over a span of 25 years. Hence, this condensed bibliography—

Alic, Margaret. HYPATIA HERITAGE, Beacon Press, Boston, 1986.

Asimov, Isaac. ASIMOV ON NUMBERS,295, Pocket Books, New York,1978.

Bakst, Aaron. MATHEMATICS ITS MAGIC & MASTERY, D. Van Nostrand Co., New York, 1952.

Ball W.W.Rouse and Coxeter, H.S.M. MATHEMATICAL RECREATIONS & ESSAYS, 13th ed., Dover Publications, Inc., New York, 1973.

Ball, W.W.Rouse. A SHORT ACCOUNT OF THE HISTORY OF MATHEMATICS, Dover Publications, Inc., New York, 1960.

Banchoff, Thomas F. BEYOND THE THIRD DIMENSION, Scientific American Library, New York, 1990.

Barnsley, Michael. FRACTALS EVERYWHERE, Academic Press, Inc., Boston, 1988.

Beckman, Petr. A HISTORY OF π, St. Martin's Press, New York, 1971.

Beiler, Albert H. RECREATIONS IN THE THEORY OF NUMBERS, Dover Publications, Inc., New York, 1964.

Bell, E.T. MATHEMATICS QUEEN & SERVANT OF SCIENCE, McGraw-Hill Book Co., Inc., New York, 1951.

Bell, E.T. MEN OF MATHEMATICS, Simon & Schuster, New York, 1965.

Bell, R.C. BOARD AND TABLE GAMES FROM MANY CIVILIZATIONS Dover Publications, Inc., New York, 1979.

Bell, R.C. OLD BOARD GAMES, Shire Publications Ltd., Bucks, U.K., 1980.

Benjamin Bold. FAMOUS PROBLEMS OF GEOMETRY & HOW TO SOLVE THEM, Dover, Publications, Inc, New York, 1969.

Bergamini, David. MATHEMATICS, Time Inc., New York, 1963.

Boyer, Carl B. A HISTORY OF MATHEMATICS, Princeton University Press, Princeton, 1985.

Brooke, Maxey. COIN GAMES & PUZZLES, Dover Publications, Inc., New York, 1963.

Bunch, Bryan H. MATHEMATICAL FALLACIES & PARADOXES, Van Nostrand Reinhold Co., New York, 1982.

Campbell, Douglas and Higgins, John C. MATHEMATICS–PEOPLE, PROBLEMS, RESULTS, 3 volumes Wadsworth International, Belmont 1984

Chadwick, John. READING THE PAST—LINEAR B AND RELATED SCRIPTS, University of California Press, Berkeley, 1987.

Clark, Frank. CONTEMPORARY MATH, Franklin Watts, Inc., New York, 1964.

Cook, Theodore Andrea. THE CURVES OF LIFE, Dover Publications, Inc., New York,1979.

Davis, Philip J. and Hersh, Reuben. THE MATHEMATICAL EXPERIENCE, Houghton Mifflin Co., Boston, 1981.

Delft, Pieter van and Botermans, Jack. CREATIVE PUZZLES OF THE WORLD, Harry N. Abrams, Inc., New York, 1978.

Doczi, György. THE POWER OF LIMITS, Shambhala Publications, Boulder, CO, 1981.

Edwards, Edward B. PATTERN AND DESIGN WITH DYNAMIC SYMMETRY Dover Publications, Inc., New York, 1967.

Ellis, Keith. NUMBER POWER, St. Martin's Press New York, 1978.

Emmet, E.R. PUZZLES FOR PLEASURE, Bell Publishing Co., New York, 1972.

Engel, Peter. FOLDING THE UNIVERSE, Vintage Books, New York, 1989.

Ernst, Bruno. THE MAGIC MIRROR OF M.C.ESCHER, Ballantine Books, New York,1976.

Eves, Howard W. IN MATHEMATICAL CIRCLES, two volumes, Prindle, Weber & Schmidt, Inc., Boston 1969.

Filipiak, ANthony S. MATHEMATICAL PUZZLES Bell Publishing co., New York, 1978.

Fixx, James. GAMES FOR THE SUPER INTELLIGENT, Doubleday & Co., Inc., New York, 1972.

Fixx, James. MORE GAMES FOR THE SUPER-INTELLIGENT, Doubleday & Co., Inc., New York, 1972.

Gamow, George. ONE, TWO, THREE . . . INFINITY, Viking Press, New York, 1947.

Gardner, Martin. PERPLEXING PUZZLES & TANTALIZING TEASERS Dover Publications, Inc., New York, 1977.

Gardner, Martin. THE UNEXPECTED HANGING, Simon & Schuster, Inc., New York,1969.

Gardner, Martin. CODES, CIPHERS & SECRET WRITING, Dover Publications, Inc., New York, 1972.

Gardner, Martin. MATHEMATICS MAGIC & MYSTERY, Dover Publications, Inc., New York, 1956.

Gardner, Martin. NEW MATHEMATICAL DIVERSIONS FROM SCIENTIFIC AMERICAN, Simon & Schuster, New York, 1966.

Gardner, Martin. MATHEMATICAL CARNIVAL Alfred A. Knopf, New York, 1975.

Gardner, Martin. MATHEMATICAL MAGIC SHOW Alfred A. Knopf, New York, 1977.

Gardner, Martin. MATHEMATICAL CIRCUS, Alfred A. Knopf, New York, 1979.

Gardner, Martin. MARTIN GARDNER'S SIXTH BOOK OF MATHEMATICAL DIVERSIONS FROM SCIENTIFIC AMERICAN, University of Chicago Press, Chicago, 1983.

Gardner, Martin. THE NEW AMBIDEXTROUS UNIVERSE, W.H. Freeman, New York, 1990

Gardner, Martin. PENROSE TILES TO TRAPDOOR CIPHERS, W.H.Freeman & Co., New York, 1988.

Gardner, Martin. KNOTTED DOUGHNUTS & OTHER MATHEMATICAL ENTERTAINMENTS, W.H.Freeman & Co., New York, 1986.

Gardner, Martin. TIME TRAVEL, W.H.Freeman & Co., New York, 1987.

Gardner, Martin. WHEELS, LIFE AND OTHER MATHEMATICAL AMUSEMENTS, W.H.Freeman & Co., New York, 1983.

Gardner, Martin. THE INCREDIBLE DR. MATRIX, Charles Scribner's Sons, New York, 1976.

Gardner, Martin. THE 2ND SCIENTIFIC AMERICAN BOOK OF MATHEMATICAL PUZZLES & DIVERSIONS, Simon & Schuster, New York, 1961.

Gardner, Martin. THE SCIENTIFIC AMERICAN BOOK OF MATHEMATICAL PUZZLES & DIVERSIONS, Simon & Schuster, New York, 1959.

Ghyka, Matila. THE GEOMETRY OF ART & LIFE, Dover Publications, Inc., New York, 1977.

Gleick, James. CHAOS, Penquin Books, New York, 1987.

Glenn, William H. and Johnson, Donovan A. INVITATION TO MATHEMATICS, Doubleday & Co., Inc., Garden City, 1961.

Glenn, William H. and Johnson, Donovan A. EXPLORING MATHEMATICS ON YOUR OWN, Doubleday & Co., Inc., Garden City, 1949.

Golos, Ellery B. FOUNDATIONS OF EUCLIDEAN AND NON-EUCLIDEAN GEOMETRY, Holt, Rinehart and Winston Inc., New York, 1968.

Graham, L.A. INGENIOUS MATHEMATICAL PROBLEMS & METHODS, Dover Publications, Inc., New York, 1959.

Greenburg, Marvin Jay. EUCLIDEAN AND NON-EUCLIDEAN GEOMETRIES, W.H. Freeman & Co., New York, 1973.

Grünbaum, Branko and Shephard, G.C. TILINGS AND PATTERNS, W.H. Freeman & Co., New York, 1987.

Gunfeld, Frederic V. GAMES OF THE WORLD, Holt, Rinehart & Winston, New York, 1975.

Hambridge, Jay. THE ELEMENTS OF DYNAMIC SYMMETRY, Dover Publications, Inc., New York, 1953.

Hawkins, Gerald S. MINDSTEPS TO THE COSMOS, Harper & Row, Publishers, New York, 1983.

Heath, Royal Vale. MATH-E-MAGIC, Dover Publications, Inc., New York, 1953.

Herrick Richard, editor. THE LEWIS CARROLL BOOK, Tudor Publishing, Co., New York,1944.

Hoffman, Paul. ARCHIMEDES REVENGE, W.W. Norton & Co., New York, 1988.

Hoggatt, Verner E., Jr. FIBONACCI & LUCAS NUMBERS, Houghton Mifflin Co. Boston, 1969.

Hollingdale, Stuart. MAKERS OF MATHEMATICS, Penquin Books, London, 1989.

Hunter, J.A.H. and Madachy, Joseph S. MATHEMATICAL DIVERSIONS, Dover Publications, Inc., NEW YORK, 1975.

Huntley, H.E. THE DIVINE PROPORTION, Dover Publications, Inc., New York, 1970.

Hyman, Anthony. CHARLES BABBAGE, Princeton University Press, Princeton, 1982.

Ifrah, George. FROM ONE TO ZERO, Viking Penquin Inc., New York, 1985.

Ivins, William M. ART & GEOMETRY Dover Publications, Inc., New York, 1946.

Jones, Madeline. THE MYSTERIOUS FLEXAGONS, Crown Publishers, Inc., New York, 1966.

Kaplan, Philip. POSERS, Harper & Row, New York, 1963.

Kaplan, Philip. MORE POSERS, Harper & Row, New York, 1964.

Kasner, Edward & Newman, James. MATHEMATICS AND THE IMAGINATION, Simon & Schuster, New York, 1940.

Kim, Scott. INVERSIONS, W.H. Freeman and Co., New York, 1981.

Kline, Morris. MATHEMATICS AND THE PHYSICAL WORLD, Thomas Y. Crowell Co., New York, 1959.

Kline, Morris. MATHEMATICS-THE LOSS OF CERTAINTY, Oxford University Press, New York, 1980.

Kline, Morris. MATHEMATICAL THOUGHT FROM ANCIENT TO MODERN TIMES, 3 volumes Oxford University Press, New York, 1972.

Kraitchik, Maurice. MATHEMATICAL RECREATIONS, Dover Publications, Inc., New York, 1953.

Lamb, Sydney. MATHEMATICAL GAMES PUZZLES & FALLACIES, Raco Publishing Co., Inc., New York, 1977.

Lang, Robert. THE COMPLETE BOOK OF ORIGAMI, Dover Publications, Inc., New York, 1988.

Leapfrogs. CURVES, Leapfrogs, Cambridge, 1982.

Linn, Charles F., editor. THE AGES OF MATHEMATICS, 4 volumes, Doubleday & Co., New York, 1977.

Locker, J.L., editor. M.C.ESCHER, Harry N. Abrams, Inc., New York, 1982.

Loyd, Sam. THE EIGHTH BOOK OF TAN, Dover Publications, Inc., New York, 1968.

Loyd, Sam. CYCLOPEDIA OF PUZZLES, The Morningside Press, New York, 1914.

Luckiesh, M. VISUAL ILLUSIONS, Dover Publications, Inc., New York, 1965.

Madachy, Joseph S. MADACHY'S MATHEMATICAL RECREATIONS, Dover Publications, Inc., New York, 1979..

McLoughlin Bros. THE MAGIC MIRROR, Dover Publications, Inc., New York, 1979.

Menninger, K.W. MATHEMATICS IN YOUR WORLD, Viking Press, New York, 1962.

Montroll, John. ORIGAMI FOR THE ENTHUSIAST, Dover Publications, Inc., New York, 1979.

Moran, Jim. THE WONDEROUS WORLD OF MAGIC SQUARES, Vintage Books, New York, 1982.

Neugebauer, O. THE EXACT SCIENCES IN ANTIQUITY, Dover Publications, Inc., New York, 1969.

Newman, James. THE WORLD OF MATHEMATICS, 4 volumes, Simon & Schuster, New York, 1956.

Ogilvy, C. Stanley and Anderson, John T. EXCURSIONS IN NUMBER THEORY, Dover Publications, Inc., New York, 1966.

Ogilvy, Stanley C. and Anderson, John T. EXCURSIONS IN NUMBER THEORY, Oxford University Press, New York 1966.

Osen, Lynn, M. WOMEN IN MATHEMATICS, The MIT Press, Cambridge, 1984.

Peat, F. David. SUPERSTRINGS AND THE SEARCH FOR THE THEORY OF EVERYTHING, Contemporary Books, Chicago, 1988,

Pedoe, Dan. GEOMETRY AND THE VISUAL ARTS, Dover Publications, Inc., New York, 1976.

Perl, Teri. MATH EQUALS Addison-Wesley Publishing Co., Menlo Park, 1978.

Peterson, Ivars. THE MATHEMATICAL TOURIST, W.H. Freeman & Co., New York, 1988.

Pickover, Clifford, A. COMPUTERS, PATTERN, CHAOS & BEAUTY,St. Martin's Press, New York, 1990.

Ransom, William R. FAMOUS GEOMETRIES, J. Weston Walch, Portland, 1959.

Ransom, William R. CAN & CAN'T IN GEOMETRY, J. Weston Walch, Portland, 1960.

Rodgers, James T. STORY OF MATHEMATICS FOR YOUNG PEOPLE, Pantheon Books, New York, 1966.

Rosenberg, Nancy. HOW TO ENJOY MATHEMATICS WITH YOUR CHILD, Stein & Day, New York, 1970.

Rucker, Rudolf V. B. GEOMETRY, RELATIVITY & THE FOURTH DIMENSION, Dover Publications Inc., New York 1977.

Sackson, Sid. A GAMUT OF GAMES, Pantheon Books, New York, 1982.

Schattschneider, Doris and Walker, Wallace. M.C. ESCHER KALEIDOCYCLES, Tarquin Publications, Norfolk, U.K., 1978.

Science Universe Series, MEASURING & COMPUTING, Arco Publishing, Inc., New York, 1984.

Sharp, Richard and Piggott, John, editors. THE BOOK OF GAMES Galahad Books New York City, 1977.

Smith, David Eugene, HISTORY OF MATHEMATICS, 2 volumes, Dover Publications, Inc., New York, 1953.

Steen, Lynn A., editor. FOR ALL PRACTICAL PURPOSES, INTRO. TO CONTEMPORARY MATHEMATICS, W.H. Freeman & Co., New York, 1988.

Steen, Lynn Authur, ed. MATHEMATICS TODAY, Vintage Books, New York, 1980.

Stevens, Peter S., PATTERNS IN NATURE, Liitle, Brown and Co., Boston, 1974.

Stokes, William T. NOTABLE NUMBERS, Stokes Publishing, Los Altos, 1986.

Storme, Peter and Stryfe, Paul. HOW TO TORTURE YOUR FRIENDS, Simon & Schuster, New York, 1941.

Struik, Dirk J. A CONCISE HISTORY OF MATHEMATICS, Dover Publications, Inc., New York, 1967.

Waerden, B.L.van der. SCIENCE AWAKENING, Science Editions, New York, 1963.

Weyl, Hermann. SYMMETRY, Princeton University Press, Princeton, 1952.

The topics from *More Joy of Mathematics* are cross indexed with the topics from *The Joy of Mathematics*. JM/ indicates pages from *The Joy of Mathematics.*

INDEX

INDEX

INDEX

INDEX

INDEX

INDEX

INDEX

INDEX